Unlocking the Moon's Secrets

Unlocking the Moon's Secrets

Secrets

From Galileo to Giant Impact

JAMES LAWRENCE POWELL

OXFORD
UNIVERSITY PRESS

Oxford University Press is a department of the University of Oxford. It furthers
the University's objective of excellence in research, scholarship, and education
by publishing worldwide. Oxford is a registered trade mark of Oxford University
Press in the UK and certain other countries.

Published in the United States of America by Oxford University Press
198 Madison Avenue, New York, NY 10016, United States of America.

CIP data is on file at the Library of Congress

ISBN 978–0–19–769486–2

DOI: 10.1093/oso/9780197694862.001.0001

Printed by Sheridan Books, Inc., United States of America

To Kemar, Amelia, Sophie, and Lawrence William

Contents

Introduction

Tranquil, beautiful, and enigmatic, our Moon has inspired artists, aided early calendar makers, told farmers when to plant, and steered seafarers to their destinations. What more could we ask of a neighbor?

Unlike other objects in the sky, we can see the Moon's most prominent surface features with only our naked eye. This allows us to note the most obvious fact about the Moon beyond its existence: it always keeps the same face turned toward the Earth. But why? Doesn't the Moon rotate around its axis, like all the other planets and satellites of the solar system? And what are those large light and dark splotches that we call the "Man in the Moon"? Every culture seems to have come up with its own imagery to describe them. The thirteenth-century Italian poet Dante Alighieri wrote of the biblical Cain bearing a bundle of thorns on his back. In his *Dialogue*, Galileo Galilei's interlocutor compared the Moon's face to the muzzle of a lion. Traditional Japanese culture saw a rabbit, people in India saw a handprint, and so on. Down-to-earth scientists noted that the Moon's light and dark regions might be analogues of the high continents and low ocean basins of the Earth, which could mean that these contrasting features had the same origin on both bodies.

Once scientists had begun to observe the Moon using primitive telescopes, they saw circular features that they named craters, after the Greek *kratēr* for "mixing bowl." This was a purely descriptive term, but accounts of volcanoes erupting within lunar craters led generations of scientists to believe that they have the same volcanic origin as the craters we see on the Earth. But any such claims were suspect, as early observers (using telescopes) also reported asphalt, coal dust, coral reefs, dust storms, fog, glaciers, ice and snow, lakes, people (Lunites or Selenites), vegetation, and atomic bomb craters. They described cities, roads, and other evidence of civilization, some of which uncannily resembled the viewers' earthly home. The use of continually improving telescopes and mapping of the Moon's surface eventually dispelled these fanciful notions.

Unlocking the Moon's Secrets. James Lawrence Powell, Oxford University Press. © James Lawrence Powell 2023.
DOI: 10.1093/oso/9780197694862.003.0001

The most fundamental question about the Moon is where it came from. Given the nearness of the Moon and the Earth in space, it seems obvious that the origins of the two bodies must be connected. This offers an opportunity. To investigate the origin of our own planet, we would naturally look to see what the oldest rocks can tell us. But erosion and the movement of tectonic plates have long since removed them, so that the most ancient rocks on the Earth are hundreds of millions of years younger than the Earth itself. In contrast, the Moon lacks the air and water to cause erosion and has too little internal heat to drive tectonic plate movement. This leaves it a primordial object, a fossil from the earliest history of the solar system that can teach us about the origin of our own planet.

Despite appearing to be an open book, the Moon has been a reluctant tutor, carefully guarding its secrets. To discover the cause of the Moon's craters (the subject of Part I of this book) took scientists three and a half centuries from the time Galileo first gazed upon it through his self-made telescope. The Apollo missions of the 1960s and '70s mounted by the National Aeronautics and Space Administration (NASA) were intended to solve the mystery of the Moon's origin (the subject of Part II), but more than fifty years later, scientists are only now closing in on the answer.

The overarching theme of this book is how science develops, complete with misunderstandings, contentious arguments, difficult-to-relinquish assumptions, and shifting views as new facts come to light. We see how new data can eventually dislodge even the most widely accepted scientific theory and how its replacement can open new areas of research and keep science moving forward. We see that scientists, in their quest to understand, never give up.

PART I
THE ORIGIN OF THE MOON'S CRATERS

1

The Ancient Astronomers

From prehistory onward, the brilliant night sky dominated the lives of our ancestors. As a young boy in the 1940s visiting my grandparents' home in the hills of Kentucky, before the arrival of rural electricity, I saw what people had seen from time immemorial: an intensely black night sky filled by a myriad of stars, with a bright Milky Way slashing across it. Today, if you can get far enough away from the light pollution of the modern world, you can still see what our ancestors saw—and marvel.

Given our innate curiosity and the practical uses of celestial phenomena for calendar making and the prediction of eclipses, it is no wonder that many ancient civilizations studied the heavens. Our modern science of astronomy is the product of a continuous tradition more than thirty-five hundred years old that involves multiple cultures and languages. Many important discoveries were made independently by different civilizations that in the ancient world were isolated from one another.

The Babylonians were the first to understand that some heavenly phenomena are periodic and the first to apply mathematics to what they saw. They kept careful records and described the appearance of Halley's Comet in 164 BCE.

In the fifth century CE, Indian astronomer Aryabhata wrote a magnificent treatise on astronomy in Sanskrit. Although his model had the Sun at the center of the solar system, he deduced that the Earth rotates on its axis and that the Moon and the planets shine by light reflected from the Sun. Aryabhata explained why eclipses occur and calculated the length of the year at 365 days, 6 hours, 12 minutes, and 30 seconds. The modern value is 365 days, 6 hours, 9 minutes, and 10 seconds.

Mayan civilization lasted from about 2000 BCE to the Spanish Conquest in the 1500s CE. The Mayans had writing and a positional number system that allowed them to create elaborate written calendars. The solar year as calculated by the Mayan calendar was more accurate than that of the Julian calendar in use at the time in Europe. The Mayans aligned their buildings

Unlocking the Moon's Secrets. James Lawrence Powell, Oxford University Press. © James Lawrence Powell 2023.
DOI: 10.1093/oso/9780197694862.003.0002

along astronomical paths and used deep wells to record the passage of the Sun along its zenith.

The Egyptians had one of the most advanced ancient civilizations, aided by their use of hieroglyphs. They built the Great Pyramid of Giza to align with the North Star at the time, not Polaris but a star called Thuban. (The Earth's axis wobbles like a spinning top in a 25,000-year cycle, pointing to different "North Stars" along the way.) The temple of Amun-Re at Karnak pointed to the position of the midwinter sunrise. At the Nabta Playa archaeological site, which dates to the fifth millennium BCE, stones were arranged in such a way that they may have been used to locate the position on the horizon of sunrise at the summer solstice. Egyptian astronomy gave rise to the greatest ancient astronomer, Ptolemy (90–168 BCE). His *Almagest* canonized the Earth-centered view of the solar system and his astronomy held sway until the advent of Nicolaus Copernicus and his Sun-centered solar system in the sixteenth century. The *Almagest* included tables that could be used to calculate the locations of the Sun, Moon, and planets; to determine when certain stars would rise and set; and to predict when eclipses of the Sun and the Moon would occur. This made the book especially useful to astronomers and helps to account for its long life.

After the Muslims conquered Egypt in 646 BCE, between the ninth and thirteenth centuries, Islamic astronomy ascended to its golden age, as reflected in the Arabic names we still use for stars such as Aldebaran and in the words "azimuth," "nadir," and "zenith," for example. The Islamic astronomers produced a voluminous literature. Alfraganus, as he was known in the West, was born sometime in the ninth century CE. He wrote a highly influential textbook that explained Ptolemy's *Almagest* in non-mathematical terms and included revised data. Other Islamic scholars expressed doubts about Ptolemy's Sun-centered model but did not go so far as to renounce it. Abu-Mahmud Khojandi built a huge sextant near today's Tehran and used it to measure the average tilt of the Earth's axis. He got 23° 32'—very close to today's average value of 23° 26'. Another Muslim scholar, Abd al-Rahman al-Sufi, was the first to describe the Andromeda Galaxy. Astronomy was only one of many sciences to flourish in the Islamic golden age.

Astronomy in China began in the Bronze Age of the Shang dynasty, from 1600 to around 1046 BCE. The ancient Chinese inscribed the shoulder blades of oxen and the undersides of turtle shells with symbols and used them for divining heavenly intent. Such "oracle bones" found at Anyang in Henan province display the Chinese names for stars and show their arrangement in the "twenty-eight mansions," or constellations. These oracle-bone inscriptions include thousands of Chinese characters and represent the earliest form of Chinese writing.

Like the Mayans, the ancient Chinese were mainly interested in using astronomical observations to make calendars. They employed these to set the dates of the most important Chinese holidays. With the rise of each new dynasty, Chinese astronomers would make new observations and prepare a new calendar. Astronomical phenomena were regarded as mandates of heaven that governed all human affairs. During the Zhou dynasty, which followed the Shang dynasty, the Chinese developed a sophisticated "luni-solar" calendar that showed both the phase of the Moon and the time in the solar year, as well as the positions of the five planets known at the time. Like our calendar today, it required the use of leap years.

Chinese astronomers were especially on the lookout for new arrivals in the heavens, such as supernovae (stars that explode catastrophically) and comets. They recorded the appearance of a supernova in the Crab Nebula in 1054 CE. Chinese scholars were the first to make a reliable record of a total eclipse of the Sun (in 780 BCE) and made the first account of the passage of Halley's Comet in 239 BCE. Modern astronomers use the long Chinese record of such events to calibrate their timescales.

Ancient Chinese astronomers invented instruments to help with their observations. One was the gnomon, of which the most familiar type is the triangular blade of a sundial. The first known gnomon is a painted stick from a Chinese site that dates to 2300 BCE. Another had a pinhole that projected a bright circle, allowing it to be used to record both the time of day and the year. In the fourth century BCE, Chinese astronomers invented the armillary sphere, or astrolabe, shown in Figure 1.1. It is made up of a skeletal set of interlocking rings that represent lines of celestial latitude and longitude, with the Sun or the Earth at the center. The Greeks independently invented it in the third century BCE. In the Han dynasty, Zhang Heng (78–139 BCE) built an armillary sphere powered by hydraulics (using water.)

As early as the fourth century BCE, Chinese scholars knew that the Moon and planets have no light of their own but instead reflect that of the Sun. They had deduced the cause of solar eclipses, and one scholar wrote instructions for predicting them. In the Song dynasty, the director of the Astronomical Observatory asked astronomer Shen Kua (1031–1095 CE) whether the Sun and the Moon resemble round balls or flat disks. Shen explained how the phases of the Moon prove that the planets and moons are round and not flat.

Figure 1.1. An armillary sphere exhibited at Beijing Ancient Observatory, a replica of one from the Ming dynasty (1368–1644 CE). A round sphere above a square base stands on a tortoise and is supported from its four corners by dragons. (Wikimedia Commons.)

But then the director asked, "Since the Sun and Moon are in conjunction and in opposition once a day, why do they have eclipses only occasionally?" This same question occurs today to many who observe an eclipse. Since the Moon orbits the Earth and the Earth orbits the Sun, why don't we have an eclipse with each new Moon, when the Moon is between the Earth and the Sun, and with each full Moon, when the Earth is between the Moon and the Sun?

Shen's answer revealed a truly remarkable understanding of the Moon's orbit relative to the Earth's.[1] From the rarity of eclipses, he deduced that the plane of the Moon's orbit around the Sun differs from the plane of the Earth's orbit, called the ecliptic, as illustrated in Figure 1.2. The term "ecliptic" came about when ancient astronomers noted that eclipses only occur when the Moon is crossing this plane. Because the two orbital planes are tilted from each other (by 5.14 degrees), eclipses occur only now and again but predictably. This small difference in orbital tilt would later loom large as different theories for the origin of the Moon failed to explain it.

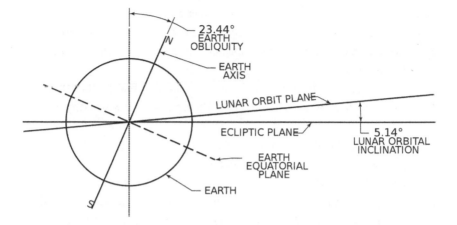

Figure 1.2. The plane of the Moon's orbit is tilted 5.14 degrees to the ecliptic; otherwise, eclipses would occur with each new and full moon. (Wikimedia Commons.)

Chinese astronomy benefited from fertilization by both Indian and Islamic scholarship. Dozens of Indian works, including some by Aryabhata, were translated into Chinese. Islamic astronomers came to China during the Mongol Empire and the subsequent Yuan dynasty; in 1210 CE, a Chinese scholar accompanied Genghis Khan to Persia to study the Islamic calendar. Books on Islamic astronomy were translated into Chinese, and many Islamic instruments were brought to China.

The Jesuit missionaries, who arrived in the late sixteenth and early seventeenth centuries, were the next to influence Chinese astronomy. Two major events in Western astronomy occurred in this period and were well known to the Jesuits: the invention of the telescope and its quick adoption by Galileo Galilei; and the introduction by Copernicus of the heretical Sun-centered theory of the solar system. The Chinese learned of the telescope from a 1615 book, and in 1634, a colleague of Galileo's presented one to the emperor. But the Chinese did not immediately adopt the telescope. Nor did they endorse and promote Copernicus's theory, as Galileo had done and for which the Church had denounced him and placed him under house arrest. The baleful Jesuit influence helped to delay Chinese acceptance of a Sun-centered solar system until the early nineteenth century.

The ancient Greeks independently made many of these same observations and went further than any ancient civilization. They noticed that among the stars, a handful of bright objects moved over a period of months and years, in repeating cycles. The Greeks named them *planētai*, meaning "wanderers." They were influenced by Babylonian and Egyptian astronomy and in turn influenced Indian astronomy.

Anaximander (ca. 610–546 BCE) was the first to realize that the Earth floats freely with nothing to hold it aloft. The philosopher Karl Popper wrote:

> In my opinion this idea of Anaximander's is one of the boldest, most revolutionary, and most portentous ideas in the whole history of human thought. It made possible the theories of Aristarchus and Copernicus. But the step taken by Anaximander was even more difficult and audacious than the one taken by Aristarchus and Copernicus. To envisage the earth as freely poised in mid-space, and to say, "that it remains motionless because of its equidistance or equilibrium" (as Aristotle paraphrases Anaximander), is to anticipate to some extent even Newton's idea of immaterial and invisible gravitational forces.[2]

Anaxagoras (ca. 500–428 BCE) believed that the heavenly bodies were stones, not divinities, so that when a meteorite fell from the sky above Thrace, some said that he had predicted it.[3] In one of his writings, he said, "It is the sun that puts brightness into the moon."[4] This allowed him to deduce that solar eclipses occur when the Moon is between the Earth and the Sun. However, by claiming that both the Sun and the Moon are "mere matter" and not gods, Anaxagoras went too far for the guardians of truth in his era and was imprisoned.

Long before Copernicus, Philolaus of Croton (ca. 470–385 BCE) considered the Earth to be a planet and asserted that it was not at the center of the universe. However, he believed that the Earth orbited a central fire of the cosmos rather than the Sun.

In the fifth century BCE, Parmenides of Elea, a disciple of Pythagoras, proposed on aesthetic grounds that the Earth is a sphere. Aristotle (384–322 BCE) confirmed this deduction by noting that the Earth's shadow during a lunar eclipse is always circular. That meant that the Earth has a circumference and therefore is subject to the rules of geometry. From this, Eratosthenes (ca. 276–195 BCE) was able to calculate the Earth's size. He had heard that on the day of the summer solstice in the ancient city of Syene (now Aswan, Egypt), the midday Sun shone directly down a well to illuminate its bottom and cast no shadow on the ground above. But on the day of the solstice in Alexandria, the noonday Sun was not quite directly overhead, casting a shadow equal to one-fiftieth of the circumference of a circle. The distance between the two cities was known to be five thousand stadia, so that the circumference of the entire circle—the Earth—was 50 × 5000 or 250,000 stadia. Eratosthenes used the Olympic stade of 184.8 meters, which multiplied by 250,000 gave a circumference for the Earth of 46,200 kilometers. This differs by a mere 15 percent from the modern figure of 40,075 kilometers measured around the equator.

Aristarchus (ca. 310–230 BCE) put together deductions of two earlier observers, Philolaus and Heracleides, to conclude that the Earth rotates around an axis and revolves around the Sun. His most astounding feat may have been his calculation of the Moon's size. By noting how long the Moon takes to travel through the Earth's shadow in an eclipse, he determined that the diameter of the Moon is roughly one-third the diameter of the Earth. The actual ratio turns out to be about 28 percent, so Aristarchus was not far off. He also estimated the size of the Sun and, long before Shen Kuo, deduced that the Moon's orbit is tilted to the ecliptic.

In the second century BCE, Hipparchus (190–120 BCE), the inventor of trigonometry, discovered that the Earth's axis wobbles over time, known as "the precession of the equinoxes." He measured the celestial longitude (the position in the sky) of certain bright, fixed stars and, when comparing them to earlier measurements, found that they appeared to have moved, leading him to discover precession. The written works of Hipparchus have not survived, but Ptolemy mentions his results. Likely, both thought that it was the heavens and not the Earth that had moved.

This summary of the remarkable deductions of the ancients testifies to the power of observation and human intellect, but ultimately, ingenious reasoning and the naked eye could take them no further.

2

Pioneers of Modern Astronomy

When Galileo Galilei (1564–1642) learned that a Dutchman had invented an instrument that could make distant objects seem near, the great Italian "gave himself up . . . to inquire into the principle of the telescope." After "deep study of the theory of Refraction," in 1609, he built his own instrument.[1] It magnified by only 3x, but that was enough to allow him to use it to observe features of the Moon undetectable by the naked eye. Galileo went on to make fundamental discoveries about the Moon and later about other satellites and planets of the solar system. These and his many fundamental discoveries in astronomy, physics, engineering, and so on through a long list, prompted former British prime minister Isaac Disraeli to call Galileo the "father of science."[2]

Galileo saw no signs of life on the Moon, writing, "I do not know nor do I suppose that herbs or plants or animals similar to ours are propagated on the moon, or that rains and winds and thunderstorms occur there as on the Earth; much less that it is inhabited by men."[3] Nor did he agree with the ancient Greeks who, starting with Aristotle, believed that the heavens consisted of concentric, solid, crystalline spheres to which the Moon and the planets were attached and which rotated in space, made of an unadulterated and perfect substance called *quintessence*. With its light and dark features visible even to the naked eye, the Moon obviously lacked perfection. Aristotle explained that its nearness to the imperfect Earth had contaminated the Moon.

Galileo also noted the Moon's irregularities: "The surface of the Moon is not perfectly smooth, free from inequalities and exactly spherical, as a large school of philosophers considers with regard to the Moon and the other heavenly bodies." On the contrary, the Moon was "full of inequalities, uneven, full of hollows and protuberances, just like the surface of the Earth itself, which is varied everywhere by lofty mountains and deep valleys."[4] He saw that the Moon was not an unknowable, perfect object but a world like our own.

As evidence of the imperfection of the Moon's surface, Galileo noted that a few days after the new moon, "When the Moon presents itself to us with bright horns, the boundary which divides the part in shadow from the

Unlocking the Moon's Secrets. James Lawrence Powell, Oxford University Press. © James Lawrence Powell 2023. DOI: 10.1093/oso/9780197694862.003.0003

shining part [which we today call the terminator] does not extend continuously in an ellipse, as would happen in the case of a perfectly spherical body." Instead, "It is marked out by an irregular, uneven, and very wavy line, as represented in the figure given" (see Figure 2.1)."[5]

Two contrasting features of the Moon especially drew Galileo's attention: the "vast prominences on the upper and lower sides of it [that] rise to an enormous elevation" and "a certain cavity larger than all the rest, and in shape perfectly round." Galileo reported his observations in a 1610 book titled *Sidereus Nuncius* and in a second work published in 1632, *Dialogue Concerning the Two Chief World Systems*. To summarize his findings in one sentence, Galileo saw on the Moon's surface a multitude of circular features of various sizes with elevated rims and flat, dark floors, sometimes containing central mountain peaks. When we come to descriptions of lunar craters

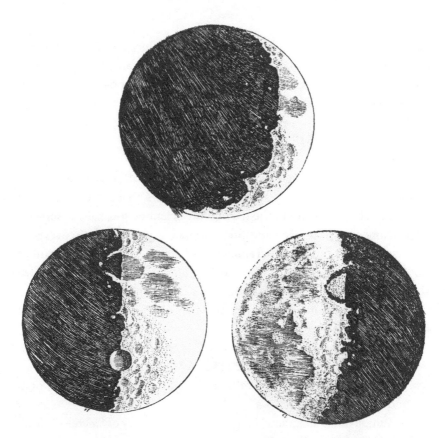

Figure 2.1. Galileo's sketches of the Moon. (Galileo, *Sidereus Nuncius*.)

made using modern telescopes and spacecraft, we will find that they largely repeat Galileo's observations, though with additional details that his instrument lacked the power to reveal.

In 1610, Galileo turned his improved, 30x telescope onto Jupiter, so much more distant than the Moon, and saw that "three little stars, small but very bright, were near the planet." As he continued to watch, a fourth swung into view. He named them the four Medicean stars after his patrons, the Medici family. Over time, he saw that these objects circled Jupiter and concluded, "It can be a matter of doubt to no one that they perform their revolutions about this planet." He also focused on Venus and observed that, like the Moon, it has a full set of phases, as Nicolaus Copernicus's 1543 theory of a Sun-centered solar system had predicted, but which Ptolemy's Earth-centered model could not explain. The phases of Venus led many early astronomers to accept Copernicus's theory.

* * *

Galileo also observed what he thought were two moons, later found to be planetary rings, encircling Saturn. He wrote to Johannes Kepler (1571–1630), to announce his finding and, using a common practice at the time, introduced the discovery as an anagram:[6]

S M A I S M R M I L M E P O E T A L E U M I B U N E N U G T T A U I R A S

Unscrambled and translated from Latin to English, this reads, "I have observed the most distant of planets to have a triple form." An anagram was used to establish priority in the days before scientific journals. If someone later claimed to have been the first to make a particular discovery, the anagrammer had only to unscramble the code and announce his primacy. But the method was not foolproof. Kepler thought the anagram (translated) should read, "Hail, twin companionship, Children of Mars," giving rise to the belief that Mars had two undiscovered moons, which it serendipitously turned out it does.

Galileo's observations led him to adopt Copernicus's theory. In 1616, however, the Spanish Inquisition ordered him to cease endorsing the theory, and when he persisted in his 1632 *Dialogue*, forced him to spend the rest of his life under house arrest.

In addition to his anagram, Galileo sent Kepler a copy of *Sidereus Nuncius* and asked him to comment on the work. Kepler had served as a mathematics teacher at a seminary school, later assisted the famous astronomer Tycho Brahe in Prague, and then had become imperial mathematician in the court of Holy Roman Emperor Rudolf II.

Like Galileo, Kepler accepted Copernicus's view that the planets orbit the Sun. But Kepler took the idea one step further. By studying the astronomical data that Brahe had meticulously collected, Kepler overturned a belief held since the Greeks: that planets travel in circular orbits. He showed that, instead, the orbits are ellipses and described the laws of planetary motion. Isaac Newton, using his own theory of gravity and his invention of calculus, was able to derive the three laws mathematically.

Kepler's response appeared in a 1634 book titled *Kepler's Conversations with Galileo's Sidereal Messenger.*[7] In contrast to Galileo, he suggested that the dark areas of the Moon's surface, the "maria," or seas, were liquid water and the bright areas were land, something that the Greek philosopher Plutarch (ca. 46–119 CE) had suggested.

Like so many others, Kepler could not resist speculating about the possibility of life on the Moon and the other planets. In *Somnium* (*Dream*), published in 1634, he gave full rein to his imagination, so much so that some regard the book as one of the earliest science-fiction novels. He explained each feature of the Moon as the work of three groups of intelligent beings. On the Moon's nearside live the Subvolvans, and on the farside reside the Privolvans, with a narrow zone in between where dwell an unnamed race of thieves. Kepler also concluded that because Galileo had to use a telescope to see the four moons of Jupiter, God must not have intended them for human eyes. But for whose, then? The residents of Jupiter, the Jovians, of course. Kepler thus was the first of what George Basalla has called "a distinguished line of astronomers, physicists, and biologists to populate the universe with beings similar to humans and societies remarkably like their own."[8]

* * *

Although Galileo accurately described the craters of the Moon, he did not discuss their cause. The first to do so was the British polymath Robert Hooke (1635–1703).[9] He became the curator of experiments at the newly established Royal Society of London, where he built telescopes more powerful than those of Galileo and Kepler and used them to make his own careful observations of the Moon. To illustrate the accuracy of these early telescopic observations, compare Figures 2.2 and 2.3, images of the Hipparchus crater. Though Hooke did not remark on it, Hipparchus illustrates another critical

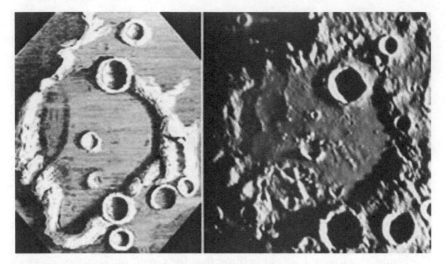

Figure 2.2. Hooke's drawing of Hipparchus crater on the left and a modern telescopic photograph on the right. (Wikimedia Commons, Lunar Photo of the Day.)

feature of lunar craters: they often cut across one another. This reveals a time sequence, showing that the crater doing the cutting came later than the one being cut. This means that the Moon, like the Earth, has a geologic history.

Hooke saw the same features that Galileo observed, especially the numerous "pits" (Hooke's word for craters). He dropped bullets into a slurry of clay and water and found that he could duplicate the appearance of the pits. But since, as far as Hooke knew, there was nothing to fall from the sky and create the pits, he thought their source must lie beneath the surface. He prepared a pot of boiling alabaster (a soft form of gypsum), which bubbled up and after it cooled displayed pits just like those he observed on the Moon. When he held a lighted candle above the alabaster in a dark room, he could reproduce all the different effects of sunlight and shadow on the Moon. Hooke concluded that the pits of the Moon and volcanoes on Earth had formed in the same way as the holes in the alabaster: by bubbling up from below. This conclusion would hold sway until the space age.

Hooke wondered, what would a person on the Moon looking back at the Earth see through a telescope? A "surface very much like that of the Moon," he concluded. Here the great scientific pioneer was wrong. When the Apollo 8 astronauts looked back at the Earth in late December 1968, they saw instead the barren, sterile Moon in the foreground, showing not the slightest

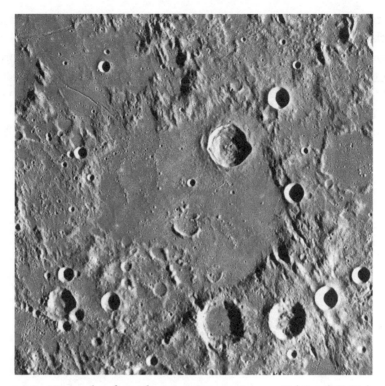

Figure 2.3. Hipparchus from the Lunar Reconnaissance Orbiter. (NASA.)

evidence of life, and, floating in space beyond, a beautiful blue and white orb, humanity's one and only home, now and, we can hope, far into the future.

Hooke also deduced that the Moon must have gravity, because, he said, "All the parts of it are so rang'd that the outmost bounds of them are equally distant from the Center of the Moon, and consequently, it is exceedingly probable also, that they are equidistant from the Center of gravitation." No one had imagined that other heavenly bodies might also have the "attractive principle," as Hooke called gravity. Then, showing proper scientific caution, he ended by calling these conjectures on the Moon "a probability, and not a demonstration," whose confirmation would have to await "future indeavours" with the telescope.

As for whether life existed on the Moon, Hooke sided with Kepler, writing that the "Vale [the floor of Hipparchus] may have Vegetables analogous to our Grass, Shrubs, and Trees; and most of these incompassing Hills may be covered with so thin a vegetable Coat, as we may observe the Hills with us to be, such as the short Sheep pasture which covers the Hills of Salisbury Plains." Such fanciful speculation should not surprise us, for those of Hooke's day, having just discovered that the Moon was not a perfect sphere but a world like our own and believing that God had populated our planet, might reasonably wonder whether he had not also endowed our heavenly companion with life. Many eminent scholars down through the centuries would describe the life they envisioned on the Moon and the planets, even on the Sun, perhaps evincing a deep human desire to believe that we are not alone in the universe.

3

Mapping and Measuring

For more than a century following Hooke's death, astronomers showed little interest in the Moon. An exception came in 1791, when German magistrate Johann Hieronymous Schröter (1745–1816) published a book titled *Selenotopographische Fragmente* (Selene was the Greek goddess of the Moon). Schröter had developed an interest in the Moon and planets through his friendship with the Herschel family of Germany, whose son William had moved to England, built his own telescopes, and became a pioneering astronomer. Schröter purchased two reflecting telescopes from William Herschel and, he said, "actuated solely by an irresistible impulse to observe," trained them first on Jupiter and then on the Moon.[1] This shift in attention was prompted by Herschel's report that he had seen three volcanoes erupting on the Moon—today we know these were the bright craters Aristarchus, Copernicus, and Kepler.[2] This false assumption bore fruit: it led Schröter to sketch the different lunar formations, or "fragments," he had observed and, using a rudimentary micrometer, to measure the heights of many lunar mountains. Galileo had been the first to make such measurements, and two other pioneering astronomers, Giovanni Battista Riccioli and Johannes Hevelius, also did so, preceding Schröter.

Galileo's method required him to recognize three things: (1) that the Moon has a topography, (2) that the bright spots in the darkness beyond the terminator are the tops of lunar mountains which catch the dying rays of the Sun, and (3) that he could use geometry to measure their heights. In *Sidereus Nuncius*, Galileo included a diagram to illustrate his method (Figure 3.1). Here we are looking at the Moon head-on, with sunlight coming from the right, illuminating the right side of the Moon while leaving the left side in darkness. Point D represents the top of a bright mountain that can just be seen by an observer to the right at G. The height of the mountain is then the distance DA. Galileo and others of his day knew the size of the Moon and thus its circumference and its diameter. This allowed Galileo to estimate the distance CD. He then knew the lengths of two sides of a right triangle—CD and CE—and from the Pythagorean theorem could calculate ED, the height

Unlocking the Moon's Secrets. James Lawrence Powell, Oxford University Press. © James Lawrence Powell 2023.
DOI: 10.1093/oso/9780197694862.003.0004

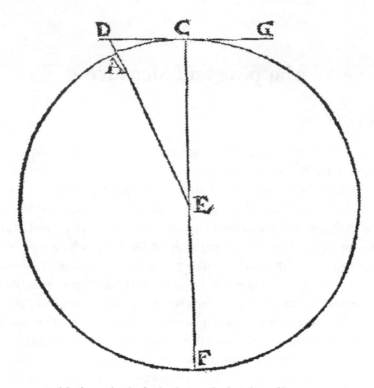

Figure 3.1. Galileo's method of calculating the heights of lunar mountains. (Galileo, *Sidereus Nuncius*.)

of the mountain plus the radius. Subtracting the radius, he had the height of the mountain.

Schröter used a different but equally ingenious method. He realized that knowing the distance from the mountain to the terminator and then calculating the altitude of the Sun above the horizon, he could use the length of the mountain's shadow to determine its height. This method had the advantage of working for any feature, including the walls of craters. You can get a sense of how it would have worked from a modern photograph of the crater Tycho, shown in Chapter 4.

Schröter also studied Venus, Mars, Jupiter, and Saturn and thus earned credit as the founder of planetology, the comparative study of planets. Like nearly everyone else at the time, Schröter believed that lunar craters were volcanoes, writing that "Nature . . . everywhere follows the same general laws, but arrived at many and diverse results."[3] Schröter was the first to apply the word "crater" to the depressions that he saw on the Moon, predisposing his successors to regard them as volcanic.[4] As for life on the Moon, Schröter joined a host of others in claiming that he had seen such evidence as green fields and a city. His mentor Herschel was so confident that life existed on the Moon that given the choice, he wrote, he would rather live on the Moon than on the Earth.[5]

William's son John would confirm the saying "like father, like son" by himself becoming a much-honored astronomer. Sir John was a true polymath who made contributions in many fields, especially early photography. In 1842, he published a "calotype"—a method of photography that used paper coated with silver iodide—of the crater Copernicus, shown in Figure 3.2. This was not a direct photograph of the crater seen through a telescope, which would not have been possible at the time. Rather, Sir John observed the crater through his own telescope, built a clay or papier-mâché model of

Figure 3.2. Photograph of the crater Copernicus using the calotype method by Sir John Herschel, 1842. (Wikimedia Commons.)

what he saw, then made a calotype of the model. This image illustrates several key features of lunar craters: near-perfect circularity, a central peak or cluster, an inner rim broken by descending terraces, small craters etched into the outer edges of the crater rim and which therefore must be younger, and the crater floor lying below the level of the surrounding area, whereas terrestrial volcanic craters are built up above the area around them. No volcanic crater on Earth comes anywhere close to presenting this full set of features.

An amusing story raises the question of whether Sir John, like his father, believed that life existed on the Moon. On Tuesday, August 25, 1835, the *New York Sun* printed the first of a six-day series of articles under this headline:

> GREAT ASTRONOMICAL DISCOVERIES
> LATELY MADE
> BY SIR JOHN HERSCHEL, L.L.D. F.R.S. &c.
> At the Cape of Good Hope

The article began in a way that should have aroused suspicion:

> In this unusual addition to our Journal, we have the happiness of making known to the British publick, and thence to the whole civilized world, recent discoveries in Astronomy which will build an imperishable monument to the age in which we live, and confer upon the present generation of the human race a proud distinction through all future time.[6]

The account continued in this overblown fashion, then gave a detailed description of the telescope, built on "an entirely new principle," as provided by the younger Herschel's "amanuensis," one Dr. Andrew Grant. Thanks to the ingenuity of the telescope designer, the instrument was so powerful that if insects existed on the Moon, it would reveal them.

Grant's account of the observations on the second day included a basalt (dark volcanic rock) shelf covered with red poppies, "that novelty, a lunar forest," a "lake or inland sea" with a beach, a "monster of a bluish lead color, about the size of a goat, with a head and beard like him, and a single horn, slightly inclined forward from the perpendicular." Capping the list would be anyone's personal favorite, a sort of winged "orang outang" given the Latin name *Vespertilio-homo*, or man-bat.

As the reader will have guessed, but many who read the original did not, these reports were entirely made up, the creation of an inventive reporter. As William Hoyt writes, whether the editor of the paper knew the reports were fake "is problematical," but he did sell a lot of papers. Sir John, who might have objected to this near-forgery under his name, is said to have enjoyed the joke.[7]

William Herschel's younger sister Caroline (1750–1848) followed in the family tradition of investigating the heavens. At age ten, she contracted typhus, which stunted her growth so that she grew to only four feet three inches and lost vision in her left eye. Her mother thought that a female so disabled need not be educated. At age twenty-two, Caroline took matters into her own hands by moving to England, where she helped manage her brother William's household. She trained in voice and gave recitals with William, who was also musical and on first moving to England had made his living by copying music. Caroline helped her brother with grinding and polishing the large mirrors that he needed for his telescopes, and she learned enough mathematics to perform the laborious calculations that his observations required. She began to use a small reflector telescope to make her own observations and discovered eight new comets. The king paid her an annual salary for her work, making her the first professional female astronomer. In 1798, she presented the Royal Society with a list of 560 stars that the *British Catalogue* had left out. Another project, a catalogue of nebulae (dark, indistinct clusters of stars), earned her the Gold Medal of the Royal Astronomical Society, which women were not allowed to join. It took 161 years for the society to award a medal to a second woman astronomer, Jocelyn Bell Burnell.[8] In 1846, two years before her death, Caroline Herschel received the Prussian Gold Medal for Science.

Another important early selenographer, German urologist and astronomer Franz von Paula Gruithuisen (1774–1852), was among the first to take up the question of the Moon's origin. In an 1828 paper, he presented the idea that the planets and their moons had all grown by the aggregation of colliding smaller bodies.[9] Unfortunately, the credibility of this prescient notion was undercut for later generations by Gruithuisen's report of people, animals, and crops on the Moon—even a fortress, a city, and a temple at which the "Lunites" worshiped the stars. These may render him foolish in our eyes, but we should remember that such delusions about lunar life were mainstream among scientists in his day.

* * *

As the early selenographers made advances in understanding the Moon's features, while coming up with ever-more-detailed speculations about what lunar life might be like, telescopes were steadily improving, making possible amazingly detailed maps of the Moon's features. One of the most famous was published in 1834–1836 by German banker and amateur astronomer Wilhelm Beer (1797–1850) and astronomer Johann Heinrich Mädler (1794–1874) (see Figure 3.3). The next year, their book *Der Mond* provided the best description of the Moon's features for its time. It included the heights of 1,095 lunar mountains and crater summits and would remain a standard reference for decades. The two also produced the first map of Mars and accurately measured its period of rotation.

The detail shown in the Beer and Mädler map of the Moon reflects countless hours of observation through the telescope. Many of the features they saw and recorded had never been seen, or at least not carefully studied, by other observers. Beer and Mädler therefore earned the right to name those

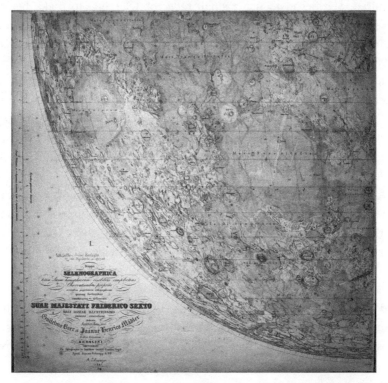

Figure 3.3 A portion of a map of the Moon's surface from Beer and Mädler, *Der Mond*. (Wikimedia Commons.)

features. Between 1876 and the end of the century, at least four other maps of the Moon appeared, some of them even more detailed.

These maps would sometimes use the same name for different lunar features and different names for the same feature. Moreover, when a feature would appear on one map but not another, it gave the false impression that the Moon was changing and thus was volcanically active. All this made it increasingly difficult for lunar scientists to communicate. British scientific organizations, including the Royal Astronomical Society, appointed an international committee to reconcile the different nomenclatures. This was a tedious task that eminent and busy astronomers would just as soon avoid, and luckily, they found just the right person to take on the job, even though her only training in astronomy had been to attend a series of university extension lectures given in the small Staffordshire town of Cheadle.

Mary Adela Blagg was born in Cheadle on May 17, 1858, and she died on April 14, 1944.[10] She did not attend college or take up a profession but rather did volunteer work and indulged in her hobby, chess. She learned mathematics from her brother's school textbooks and in 1875 attended a finishing school in Kensington to study algebra and German. Blagg had a heart condition that made it difficult for her to travel. Fortunately, at the nearby extension lectures, she caught the attention of the speaker, Joseph Hardcastle, who happened to be John Herschel's grandson. He suggested that Blagg attempt original work in his own field, selenography. Her interest in lunar nomenclature came to the notice of one of the nomenclature committee members, who invited her to assist in collating the many inconsistent names. This led to her *Collated List* published in 1913, followed by *Named Lunar Formations*, a two-volume set produced in 1935 that became the standard reference on lunar nomenclature.

At the invitation of another astronomer, Blagg began work on stars whose brightness repeatedly changed. The data were in a series of notebooks, presenting her with another tedious but important task. Again, she succeeded, coauthoring a series of ten papers. Her collaborator said, "Practically the whole of the work of editing has been undertaken by Miss Blagg. The difficulties of identification have been noted frequently; they could scarcely have been overcome without her patience and care."[11] Her entry in the *Biographical Encyclopedia of Astronomers* notes that Blagg "might have become a professional astronomer if the opportunity had presented itself. She managed to succeed in astronomy partially because she was willing to work under the direction of others and to undertake tedious problems rejected by male astronomers. Her skill and good judgment in approaching these problems assured that her contributions were more than mere fact collecting."[12]

The Royal Astronomical Society recognized Blagg's importance and elected her a fellow in 1915. After her death, most appropriately, the International Lunar Committee assigned the name Blagg to a lunar crater.

* * *

One thing that even the most meticulous map did not show was the slightest evidence of life on the Moon. In 1902, Irish astronomer Agnes M. Clerke wrote a popular, authoritative, and eloquent history of astronomy. She said that the increasingly detailed maps of the Moon "gave form against the sanguine views entertained by Hevelius, Schröter, Herschel and Gruithuisen as to the possibilities of agreeable residence on the moon, and relegated the 'Selenites'" and their "cities . . . and festal processions . . . to the shadowy land of the Ivory Gate [a reference to dreams]."[13] She was not quite right about the demise of belief in developed life on the Moon, as Harvard astronomer William Henry Pickering, would write in 1921 that he had seen plants on the Moon, sufficient for two crops a day. But by this time, he stood alone.

For the most part, by the middle of the nineteenth century, a focus on the origin of lunar craters had begun to replace speculation among selenographers about whether the Moon harbored life. With rare but important exceptions, no one showed any interest in the origin of the Moon itself. The science of astronomy was advancing rapidly, with most practitioners turning their attention outward to distant stars, which they found to be like our own Sun. But as they stepped off the stage of lunar studies, one of the most prominent geologists of the nineteenth century stepped onto it.

4

The Moon's Myriad Craters

As the early European selenographers were looking at lunar craters through their telescopes and declaring them volcanic, American geologist James Dwight Dana (1813–1895) was studying terrestrial volcanoes close-up, even witnessing them erupting, a rare event for any American geologist. After his graduation from Yale, Dana had spent a year tutoring Navy midshipmen on the USS *Delaware*. The ship called at the site of one of the world's most famous volcanoes, Italy's Vesuvius, which had erupted in 79 CE and destroyed Pompeii. Seeing Vesuvius inspired Dana's first scientific paper.[1]

Another fortunate event for Dana was an invitation to join the Wilkes Expedition of 1838–1842, which explored the Pacific Ocean. The voyage took Dana to many volcanic islands, most notably to the Hawaiian chain and Kilauea volcano on the Big Island. Dana observed the three-mile-wide summit "pit" of Kilauea, which was continually active during the nineteenth century and underwent a major eruption in 1840. He found that the lavas in the pit were so fluid that they appeared to boil. He thought this could explain the puzzling fact that all lunar craters were close to perfect circles, since "a boiling pool necessarily, by its own action, extends itself circularly around its centre."[2] He concluded that "The Moon's volcanoes are in fact volcanoes, either extinct or active, although the craters would receive comfortably more than a score of Etnas," referring to another large and active volcano in eastern Sicily. Here Dana called attention to another fact that distinguished lunar from terrestrial craters: some of those on the Moon are vastly larger and deeper than any on Earth. Hipparchus, for example, which was shown in Figure 2.3, is about 150 kilometers in diameter. In the fourth edition of his influential *Manual of Geology* in 1895, Dana, who by then had received too many honors to list and had become one of the world's leading geologists, wrote that "The principles exemplified on the earth are but repeated in her satellite,"[3] thus again ignoring the great disparity in size between lunar and terrestrial craters.

In his 1846 article "Origin of Continents,"[4] Dana noted that in the Moon's dark seas and lighter-colored highlands (see Figure 4.1), from which

Unlocking the Moon's Secrets. James Lawrence Powell, Oxford University Press. © James Lawrence Powell 2023.
DOI: 10.1093/oso/9780197694862.003.0005

Figure 4.1. Telescopic view of the full Moon, 2010. (Wikimedia Commons, Gregory Rivera.)

observers over the centuries have imagined our "Man in the Moon," he found "hints on another topic of great interest, relating to the distribution of land and water on our globe." This was to have great ramifications for the science of geology. He noted "nearly one third of the [lunar] hemisphere facing the earth, which is mostly free from volcanoes [these were the mare basins], while on other parts the craters are closely crowded together." Dana concluded that these "naked" areas had solidified first and "therefore cooled the longest and to the greatest depth" and "become somewhat depressed."

Then, reasoning by analogy from the Moon to the Earth, Dana wrote:

It is therefore a just conclusion that the areas of the surface [of the Earth] constituting the continents were first free from eruptive fires. These portions cooled first, and consequently the contraction in progress affected most [of] the other parts. The great depressions occupied by the oceans thus began; and for a long period afterward, continued deepening by slow, though it may have been unequal, progress.

This conclusion led scientists to view the Earth's ocean basins and continents as original features that have remained in place where they were first formed. "Permanence theory," as it became known, would dominate geology until the beginning of the twentieth century, when new evidence falsified it. Thus, we see that Dana reasoned by analogy from the Earth to the Moon and concluded that lunar craters were volcanic. He then analogized in the opposite direction and came up with permanence. Both analogies turned out to be false.

* * *

In his *Micrographia* of 1665, Robert Hooke identified volcanism as the probable cause of lunar craters, rather than something falling from above, in part because he knew of nothing in the sky that could have fallen to blast out a crater. As Hooke explained, "It would be difficult to imagine whence those bodies should come."[5] The recognition that such objects do exist we owe to German lawyer and polymath Ernst Florens Friedrich Chladni (1756–1827).[6] Chladni was a precocious student of mathematics and science whose father insisted that he study law and philosophy instead. Chladni dutifully did this, earning his doctorate in law at Leipzig. When his father died, Chladni was free to return to his first love: science. He wrote a book on the theory of sound waves that has led some to call him the "father of acoustics."

When the elderly physicist and philosopher Georg Lichtenberg told Chladni that he had witnessed a bright fireball fall from the sky onto Göttingen, it so piqued Chladni's interest that he spent three weeks at the library compiling a list of reported masses of iron and stone objects that had fallen from the sky throughout history. These witness accounts were so similar that Chladni regarded them as true. In a 1794 book, he proposed that what we now call meteorites arrived from space at high velocities and ignited into fireballs as they passed through the atmosphere and struck the Earth.

One of the meteorites he read about was the "Pallas Iron," a 725-kilogram mass of solid iron found near Krasnoyarsk in the mountains of Siberia and named for the German naturalist who studied it, Peter Pallas (1741–1811). It became the type example of a class of metallic meteorites known as "native irons" and later called pallasites. Local people claimed that it had fallen from the sky, and Chladni agreed, as it differed from the country rock of the region, which also lacked volcanoes and smelters to provide alternative sources for the iron mass. Another hard-to-miss meteorite was a giant iron from the Chaco region of northern Argentina, estimated to weigh 15,000 kilograms. One collector who had tried to knock off samples told of wearing out seventy chisels to get 12 kilograms of what became known as the Campo de Cielo meteorite.[7] Local legend again held that the giant iron had fallen from the sky. A modern radiocarbon date on a nearby piece of charred wood gave an age of four thousand years, so indigenous people in the area could have seen it fall, something so spectacular that it might have lived on in tribal memory.

Critics of Chladni's book argued that the alleged eyewitness accounts of meteorite falls were merely unreliable folk tales and Lichtenberg went so far as to wish aloud that Chladni had not written the book. Then, in July 1794, before the forthcoming negative reviews of the book had appeared in print, further support for Chladni's hypothesis arrived when a shower of stony meteorites fell on Siena in Tuscany, then a city of nearly thirty thousand people. Some thought a volcanic eruption had thrown the stones aloft— eruptions not on the Earth but on the Moon. The following year, observers at Wold Cottage in England witnessed the fall of a hard-to-miss, 25-kilogram stone meteorite, still the largest known ever to fall in the British Isles. In 1802, a chemical analysis showed that stony meteorites differed from terrestrial rocks especially in their higher content of the element nickel. As these meteorite falls and analyses began to give credence to the idea of rocks from space, another lucky (for Chladni) event occurred in April 1803, when some three thousand meteorites descended on the Norman town of L'Aigle.

Chladni looms large in the history of meteoritics not only because he was the first to recognize them for what they were but also because he was the first to propose that both planets and their moons had grown by the aggregation of impacting smaller bodies. Gruithuisen followed up on this idea in 1828 by developing the first model of how the mechanics of impact cratering worked on the Moon.[8] His thought ranged even further: to terrestrial analogs of large-scale lunar features, to the role of meteorite impact in the history of life on Earth, and to comets as a source of the Earth's water. But these ideas were too far ahead of their time.

* * *

In the 1870s, two of the most important nineteenth-century books on the Moon appeared: Richard Anthony Proctor's *The Moon: Her Motions, Aspect, Scenery, and Physical Condition* and James Nasmyth and James Carpenter's

The Moon: Considered as a Planet, a World, and a Satellite.[9] The two books came to opposite conclusions about the origin of the Moon's craters.

Following the loss of a family lawsuit and a bank failure, Richard Anthony Proctor (1837–1888) had found himself obliged to try to make his living by writing popular books and articles on science, of which he produced dozens. His book titles included *Easy Star Lessons, Half-Hours with the Stars, Our Place among Infinities, Easy Lessons in the Differential Calculus, Light Science for Leisure Hours*, and so on. But the popularizer of science, as even the great Carl Sagan learned, may earn the derision of colleagues. According to the *Complete Dictionary of Scientific Biography*, Proctor failed miserably: "As an astronomer he was prone to speculation, and frequently he was wildly mistaken."[10] Proctor was not alone in being consigned to the dustbin of history by this seemingly authoritative dictionary. Its entry for Chladni, considered the founder of meteoritics, begins, "Except for a few publications on meteorites, in which he proposed their extraterrestrial origin, Chladni devoted his research to the study of acoustics and vibration."[11]

Like Chladni and Gruithuisen, Proctor made a forthright case for the origin of lunar craters by meteorite impact, which he saw as fundamental to the origin of the solar system itself: "[It] is my opinion that the solar system had its birth, and long maintained its fires, under the impact and collisions of bodies gathered in from outer space."[12] He pointed to the lack of any "real resemblance between any terrestrial feature and the crateriferous surface of the moon."[13] Proctor rejected Hooke's belief that lunar craters derived from volcanism, because the Moon had craters vastly larger than any terrestrial volcano.

Proctor did acknowledge that the circularity of lunar craters was hard to explain, since incoming objects would have arrived at a variety of angles. Those that landed obliquely would presumably have left oval-shaped craters, but all those seen on the Moon's surface were circular. Proctor explained the discrepancy by noting that even after an oblique strike, the "plastic surface would close in round the place of impact" and turn an oval into a circle. Thus, the lack of oval-shaped craters was no reason to reject the impact hypothesis. Here Proctor was decades ahead of his time.

Had Proctor stopped there and never written another word, historians might remember him as one of the pioneers of planetary science. But he did not stop. Whereas we might criticize some nineteenth-century authors of science books for publishing multiple editions with little change, in the second edition of *The Moon*, written five years after the first, Proctor reversed himself and abandoned the impact theory that he had laid out so straightfor-wardly and, to a modern eye, so convincingly.

In 1881, Proctor published *The Poetry of Astronomy*.[14] It included an essay titled "The Moon's Myriad Small Craters," in which, surprisingly, he reversed himself again and returned to his original meteorite impact theory. He fo-cused on one fact he thought dispositive: "the amazing number of the lunar craters." He cited Galileo, who "with his weak telescope . . . could see but a few of the craters which really exist in the moon, compared those in the south-western part to the eyes in a peacock's tail." As telescopic power has increased, Proctor said, "more and more craters have been seen." Proctor concluded that meteorite impact had created even the smaller lunar craters, as there was no other plausible explanation. Then he drew a second prescient conclusion. "It would be a problem of extreme difficulty," he wrote, "to show how a body formed like the moon, exposed to similar conditions, and for the same enormous time-intervals, could fail to show such markings as actually exist on the moon." He did not mention the Earth as one such body, but there seems little doubt that he had it in mind.

Understandably, these alternating positions confused Proctor's fellow scientists, as well as historians of science, about what he truly believed re-garding the origin of lunar craters. Proctor may well have squandered the chance to give meteorite impact a scientific toehold, a failure that allowed the second important book of the 1870s to become the standard authority and to help delay by decades the acceptance of meteorite impact on the Moon.

* * *

Scotsman James Nasmyth (1808–1890) was an engineer and inventor who specialized in power tools and came up with the idea of a hammer driven by steam power. In retirement, he built his own 20-inch reflecting telescope and made many observations of the Moon and sketches of its features. Plaster models of the Moon's surface reproduced the effects of light and shadow on the craters. Nasmyth then took photographs of the models and included them in the book he wrote with James Carpenter (1840–1899). As shown in Figure 4.2, it worked wonderfully well. Amateur astronomers could open Nasmyth and Carpenter's book and see the Moon's surface from their armchairs, anytime they liked and regardless of weather. And if authors having such detailed evidence as shown in the photographs endorsed volcanism to explain the Moon's surface features, who was to doubt them?

PLATE VIII.—Aristotle and Eudoxus.

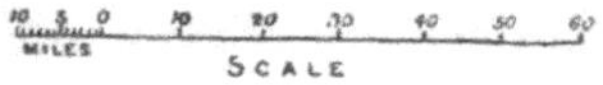

Figure 4.2. Plate VIII from Nasmyth and Carpenter, *The Moon*.

In their preface, Nasmyth and Carpenter, the latter an astronomer at the Greenwich Observatory, made it clear that they had a single aim: to present a brief for "volcanic energy" as the cause of "the characteristic craters and other eruptive phenomena that abound upon the moon's surface." They nowhere mentioned meteors except to discuss the minute ones that become "shooting stars," too small to have disturbed the Moon's surface. Nor did they cite the meteorite impact theory as proffered by Gruithuisen and later by Proctor. We might imagine that Proctor's book appeared too late for Nasmyth and Carpenter to cite it in their 1874 edition. But that does not satisfy, for the two

Figure 4.3. Tycho from the Lunar Reconnaissance Orbiter. (NASA.)

also failed to mention Proctor in their 1903 final edition. Nor in that edition did they mention an 1893 article[15] on the Moon's surface by the foremost student of topography, geologist G. K. Gilbert.

Nasmyth and Carpenter wrote that whenever a lunar crater resembled a terrestrial one, the same process had formed both. Logically, this should have meant that where the two differed, their formation could also have differed. Instead, Nasmyth and Carpenter attributed each such discrepancy to the different conditions on the Earth and the Moon.

Again, the sheer size of lunar craters should have given the two authors pause. As they noted, Tycho (see Figure 4.3) is "54 miles [87 kilometers] in diameter, and upwards of 16,000 feet [4.9 kilometers] deep, from the highest ridge of the rampart to the surface of the plateau, whence rises a grand central cone 5000 feet high." Earth has nothing comparable. Medium-sized craters like Tycho typically have a central peak, but larger ones do not. Clavius (see Figure 4.4), for example, the second-largest crater on the nearside of the

Figure 4.4. Clavius by telescope. (Wikimedia Commons.)

Moon at 231 kilometers in diameter and 3.5 kilometers deep, does not have one, but at least two craters inside Clavius do. The largest craters and basins (defined as those more than 300 kilometers in diameter), like the 1,000-kilometer-wide Orientale Basin shown in Chapter 12, lack a central peak, having instead a circular ring inside the crater. These giant "ring formations" stumped Nasmyth and Carpenter, leading them to "face the possibility" that the features had some cause other than volcanism. The two could not identify that cause and so fell back on Dana's theory that the ringed basins were once gigantic, boiling lava pits—gigantic indeed, since the entire Hawaiian Island chain could have fit inside Clavius.

Nasmyth and Carpenter could not so easily brush aside one seemingly dispositive difference between lunar and terrestrial volcanism. On the Earth, Nasmyth and Carpenter recognized, water-born steam drives volcanic eruptions, and Nasmyth was an expert on steam. But the Moon has far too little water to generate steam. To escape the dilemma, the two authors appealed to a strange and counterintuitive process called "expansion upon solidification": as rocks cool, they supposedly expand, pushing up what-ever lies above them until they finally burst through to the surface. As the single piece of evidence for this supposed process, they cited an experiment by British scientist John Tyndall, who had put molten bismuth into an iron bottle and found that as it cooled, the bismuth burst open the bottle. This lone experiment, based on the peculiar properties of an exotic substance, provided the slender reed on which Nasmyth and Carpenter hung lunar vol-canism: the more rocks cool, the more they expand, until finally they explode.

Near the end of their 1874 edition, Nasmyth and Carpenter included a chapter titled "The Moon as a World: Day and Night upon its Surface." After having spent much of the book describing lunar features that could have derived from meteorite impact yet never having considered the possibility, it comes as a surprise to find them writing, "Dark meteoric particles and masses would continually bombard the lunar surface, [striking] the moon with a force to which that of a cannon-ball striking a target is feeble indeed." Their final edition, in 1903, contains the same language. Evidently, having closed their minds to any process but volcanism, Nasmyth and Carpenter could not imagine that larger meteorites striking with a force much greater than a cannonball could account for nearly every lunar feature they describe.

In 1893, between the two editions, Gilbert presented persuasive evidence that the surface features of the Moon do indeed result from meteorite impact. Nasmyth and Carpenter never mentioned his incisive work and decades would pass before others would pay any attention to it.

5

Colliding Moonlets

Aside from the giants who wrote about the Moon before the era of scientific specialization—Galileo, Kepler, and Hooke, most notably—until the 1890s, the Moon was primarily the domain of amateur astronomers. Professionals had shown little interest, preferring to study the stars and galaxies that were becoming increasingly visible through their ever-improving telescopes. As we saw, Nasmyth (a civil engineer) and Carpenter (the exception, an astronomer) compared the surface features of the Moon with those of the Earth, especially with terrestrial volcanoes. But topographic features belong to the science of geology, not astronomy. In 1893, the first report on the topography of the Moon—the character and origin of its surface features—by a geologist appeared. And what a geologist he was.

Biographer Stephen Pyne described Grove Karl Gilbert (1843–1918) as "A Great Engine of Research," as Gilbert himself had called the US Geological Survey, where he served as chief geologist from 1889 to 1892.[1] Gilbert was the first geologist to explore the American West and became the scientific partner of another pioneer, John Wesley Powell, the one-armed former Civil War officer who became famous for his death-defying raft trip down the Colorado River and through the Grand Canyon and who later served as the second director of the Survey.[2] When Gilbert spoke, geologists listened.

No one was better qualified than Gilbert to scrutinize the Moon's surface and interpret its landforms. In 1892, he spent eighteen nights viewing the Moon through the 26-inch refractor of the US Naval Observatory in Washington, DC, supplementing his findings with lunar photographs from California's Lick Observatory, on file at the nearby Smithsonian Institution. Some did not think it appropriate for a geologist to spend his time staring into space rather than at the rocks at his feet. In an 1892 congressional discussion of the failings of the US Geological Survey, one congressman declaimed, "So useless has the Survey become that one of its most distinguished members has no better way to employ his time than to sit up all night gaping at the Moon."[3] Gilbert himself admitted to a friend, "I am a little daft on the subject of the moon."[4]

Gilbert delivered his conclusions in December 1892 as the retiring president of the Philosophical Society of Washington and presented them at several scientific meetings. In 1893, the society published his address as *"The Moon's Face: A Study of the Origin of Its Features.* Anyone familiar with the writings of nineteenth-century science authors such as Dana, Proctor, Nasmyth, and Carpenter, with their ornate phrasing, long and convoluted sentences, and overuse of the passive voice, finds pleasant relief in Gilbert's direct and terse style. The opening sentence of his published address reads, "The face which the moon turns ever toward us is a territory as large as North America, and, on the whole, it is perhaps better mapped."[5] These few words tell us three things right off the bat: the Moon is locked onto the Earth, the area of the Moon's face (the "nearside") is roughly the same as that of one of our continents, and we know more about the details of the Moon's surface than we do about those of our own planet. Even in the 1890s, Gilbert wrote, scientists knew the Moon's face so well that there was no place left on its map to write "unexplored."

Since, as he wrote, "All theories [about the Moon] begin with craters," Gilbert starts with a description of their characteristics, illustrating it with a sketch of the "type form," shown in Figure 5.1. The differences "far outweigh the resemblances" to terrestrial volcanic craters, Gilbert wrote, and could not have arisen simply from differing conditions on the two bodies.

Like Hooke, Gilbert dropped pebbles into a pool of mud and found that he could easily replicate lunar craters, confirming in his mind that they were the result of impact. Still, he wrote, the nearly perfect circularity of lunar craters presented a "most formidable difficulty" for the meteoric theory. In

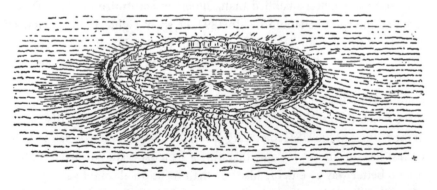

Figure 5.1. Typical lunar crater by Gilbert, showing the central peak, terraced inner rim, flat floor, and sloping outer rim. (Gilbert, *The Moon's Face.*)

his experiments, Gilbert had found that when he flung an object at the mud obliquely, the result was an oval depression. But the Moon had no oval craters. Gilbert got around this potentially fatal flaw by imagining that the impacting objects—or moonlets, as he called them—had fallen vertically, not obliquely, and at much lower velocities than typical cosmic speeds. He envisioned the moonlets as having populated a ring resembling those of Saturn, confined to a single plane that orbited the primordial Moon. Gilbert went through an elaborate analysis to show that the moonlets would have fallen to the surface vertically and left circular craters.

He went on to discuss the Moon's surface features other than craters, focusing on the white streaks such as those extending out from Tycho (see Figure 4.3), which we call rays. These "stretch for long distances across the moon's surface" and "pass up and down the slopes of craters without either modifying their forms or being interrupted by them." Each ray system radiates from a crater, leading Gilbert to conclude that they were splashes of debris thrown up by the impact of a meteorite.

Near the end of *The Moon's Face*, Gilbert asked: "Does the earth exhibit impact craters? If not, then erosion and sedimentation have destroyed them." Thus, he espoused what almost all other scientists would doubt until the 1960s: that the Earth must also have had impact craters. He went on to suggest that the Moon itself had formed from the aggregation of small bodies.

Gilbert's moonlet hypothesis attracted little interest. An exception was a review in *The Nation* by the eminent Harvard geomorphologist William Morris Davis (1850–1934), a friend and colleague of Gilbert.[6] Davis sided with his fellow geologist, noting that scientists had never observed volcanic eruptions on the Moon, so there was no actual evidence of their existence. He also pointed to problems with his colleague's thesis, such as the likelihood that at least a few lunar craters should be oval-shaped, yet there was none. Davis summed up: "The evidence from imperfect analogy on which the volcanic theory has been so long sustained needs to be carefully reviewed." In 1922, eight years after Gilbert died, Davis penned the great man's biographical memoir for the National Academy of Sciences.[7] It amounted to 303 pages at around 800 words per page, for a total of around 240,000 words. But if the career of any geologist had earned such a lengthy tribute, it was Gilbert's.

In his memorial, Davis returned to the subject of the craters of the Moon to make one particularly telling remark: "Each group of scientists find the craters so difficult to explain by processes with which they are professionally familiar that they have recourse to a process belonging in another field than their own, with which they are probably imperfectly acquainted, and with which they therefore feel freer to take liberties."

* * *

Gilbert's curiosity about impact craters had preceded his Moon-gazing. At the 1891 meeting of the American Association for the Advancement of Science, he was in the audience when mineralogist Arthur E. Foote described a "circular elevation . . . occupied by a cavity nearly three-quarters of a mile in diameter."[8] It was near the northern Arizona trading post of Canyon Diablo and was known locally, if illogically, as "Coon Mountain." Strewn around the hole lay unmistakable iron meteorites, Foote related, some containing diamonds. He also called attention to the absence of "lava, obsidian, or volcanic products."

In the discussion that followed, Gilbert said that the "so-called crater" resembled the "depressions on the surface craters of the moon produced by the impact of an enormous meteorite mass." Thus, even before his nights spent viewing the Moon by telescope, Gilbert believed that impact had created its craters and perhaps was responsible for the Coon Mountain cavity as well. Note from Figure 5.2 how well the former Coon Mountain resembles

Figure 5.2. Meteor Crater, Arizona, the former Coon Mountain. (Wikimedia Commons.)

Gilbert's sketch of a typical lunar crater in Figure 5.1, though it is not large enough to have a central peak.

Perhaps to get away from the bloviations of Washington, DC, and certainly to get back into the field, in November 1891, Gilbert took a leave of absence from the USGS and went to Arizona to see the diamondiferous crater for himself.[9] He wrote to a friend of his "peculiar errand to hunt a star," noting that "numerous fragments of meteoric iron have been found in the tract adjacent to a 'crater,' and the crater differs from others in that it is composed of sandstone and limestone and has no volcanic rock." Gilbert thought that the cavity was "the scar produced by the collision of the earth with a small star. The indications are that the missile is somewhere under the scar."

At this time, prominent geologists espoused the method of "multiple working hypotheses"—holding in mind several simultaneously, to avoid settling prematurely on a single hypothesis. Gilbert made the case this way:

> The great investigator is primarily and preeminently the man who is rich in hypotheses. The man who can produce but one cherishes and champions that one as his own, and is blind to its faults. With such men, the testing of alternative hypotheses is accomplished only through controversy. Crucial observations are warped by prejudice, and the triumph of truth is delayed.[10]

For the origin of Coon Mountain, Gilbert had two hypotheses in mind: either meteorite impact or a blast of volcanic steam from below that excavated the crater but left no lava or other telltale evidence of volcanism. He believed that by comparing the volume of material in the crater rim with the volume of the crater, he could decide between the two. Here is how this test was supposed to work, based on prevailing knowledge at the time. If an impacting meteorite had blasted out the crater, it would still be present and would take up some of the space in the resulting cavity. The presence of the meteorite would cause the crater volume to be less than the volume of material in the rim, which had been excavated from inside the crater. If, on the contrary, the volumes of rim and crater were the same, there was no room in the crater for a buried "star," falsifying the impact hypothesis. As a second test, the magnetism of a buried iron meteorite would reveal its hidden presence, though many meteorites are "stones" without unusual magnetism.

Gilbert had at first "ascrib[ed] the crater to a falling star," by which he meant a meteor, because otherwise the meteorite fragments scattered around Coon Mountain would have to be put down to coincidence, always a scientific last resort. Indeed, Gilbert estimated "[t]he probability of non-coincidence to be at least 800 times as great as the probability of coincidence," a ratio that he said "legitimately [inclines] the mind toward causality."[11]

Gilbert and his field assistant made a careful study of Coon Mountain—its topography, geology, magnetism, and so on. To his surprise, the volume of material in the rim turned out roughly to match the volume of the crater, supporting the hypothesis of a steam eruption and, by Gilbert's inexorable logic, ruling out meteorite impact. As for his second test, Gilbert found no abnormal magnetism in and around the crater and concluded that any buried meteorite would have been tiny or located at great depth. (Importantly, in neither case would it be worth trying to mine for iron ore.)

Since both tests—the matching rim and crater volumes and the absence of magnetism—had failed to support the impact hypothesis, Gilbert's unbending chain of logic demanded that he accept a finding that he had given odds of 800 to 1 against. As he explained in a letter:

> I didn't find the star of course, because she is not there. . . . The theory at present in favor is that hot lava was injected 1200 or 1500 feet below the surface of the plain, superheating some water it found there and causing a big blast of steam power. If that theory is good, the shower of meteoric iron was subsequent, and its coincidence in place was fortuitous.[12]

Gilbert did his Arizona fieldwork in 1891, published *The Moon's Face* two years later, and presented his Coon Mountain conclusions in an 1896 paper titled "The Origin of Hypotheses, Illustrated by the Discussion of a Topographic Problem," the problem being the origin of the Coon Mountain crater. He described his conclusions in talks at several important meetings and at universities, defending what must have seemed to his audiences a most illogical deduction.

Showing himself willing to entertain alternative hypotheses, Gilbert noted that another way to explain the lack of magnetism at Coon Mountain was for the impactor to have contained embedded meteorites, "like plums in an astral pudding," instead of one large magnetic object. If so, then the lack of magnetism in the crater would not necessarily falsify the impact theory. He also found a way around the apparently equal volumes of crater and rim, noting that shock pressure could have condensed the target rocks into a smaller space. (He had no way of knowing that, as we will see, a small impacting meteorite would blast itself into fragments and create a much larger crater, obviating this test.) Gilbert wrote that no conclusion, "[h]owever grand, however widely accepted, however useful, is so sure that it cannot be called in question by a newly discovered fact. In the domain of the world's knowledge there is no infallibility."[13] Given this seeming change of mind, we might have expected Gilbert to declare the origin of the Coon Mountain crater undecided pending new evidence. But Gilbert failed to live up to his noble principles, never writing another word about the crater. He held such a commanding position in American geology and at the Geological Survey, not just during his lifetime but long after, that when he declared that Coon Mountain was not an impact crater and never spoke or wrote of the matter again, impact cratering became off limits to American geologists and remained so until the 1960s. Gilbert was wrong about the origin of Coon Mountain, but no one challenged his conclusion. He was right about the origin of lunar craters, and nearly everyone disagreed or ignored him.[14] Not until 1959 did any Survey geologist contradict Gilbert in print. That geologist was the pioneer of modern meteorite impact studies, Eugene Merle Shoemaker (1928–1997). In 1983, and fittingly, Shoemaker won the inaugural G. K. Gilbert Award of the Geological Society of America. The next year, he won the first Barringer Medal of the Meteoritical Society.

* * *

When in 1902 mining engineer Daniel Moreau Barringer (1860–1929) learned of the Coon Mountain crater, he developed a lifelong conviction that despite Gilbert's conclusion, underneath the cavity lay a valuable and mineable mass of extraterrestrial iron ore. Barringer acquired the rights to mine the crater and wrote letters by the hundreds to arouse interest in his project (and boost his company's stock price). In 1932, the US Board on Geographical Names changed the name of Coon Mountain to Crater Mound. Then, in 1946, the Board changed the name again, this time to Meteor Crater, no doubt to the delight of the Barringer family.

To be worth mining, the buried meteorite had to be of sufficient size. Based on the dimensions of the crater, Barringer had estimated its mass at 100 million tons. Had it consisted of solid iron, like Pallas and many meteorites, at the then-current (1903) price of $20 per ton of iron, the missing meteorite would have been worth $2 billion in today's dollars, a fortune well worth obsessing over.

Investors in the project asked astronomer Forest Ray Moulton (1872–1952) to prepare a definitive report on the size of the putative impactor.

Using the equation for the energy of a moving object, Moulton calculated that the meteorite would not have remained intact but would have broken into fragments that collectively would have weighed 300,000 tons, far less than Barringer's estimate. A confident Moulton joked, "If more than 500,000 tons of meteorite are found, take a long and hearty laugh at my expense and call on me to put up a dinner at the crater for all interested. If my reasoning and my results are so much in error, I will come across with the dinner."[15]

Not surprisingly, the Barringer family vehemently objected to Moulton's determination of a relatively puny, fragmented, and therefore nearly worthless meteorite. Barringer himself continued a voluminous correspondence on the crater, his last letter written on November 27, 1929. Two days later, he suffered a heart attack and died. His obituary in the New York Times, well ahead of the formal name change, noted that Barringer had "discovered the origin of the famous Meteor Crater of Arizona and proved that it is due to the impact of a meteorite mass." No doubt, the USGS disagreed. Ironically, at a 2018 auction, a 70-pound fragment of the Canyon Diablo meteorite from Meteor Crater brought $237,500. Another specimen played a key role in the 1953 discovery of the age of the Earth and the solar system. On that, it would be hard to put a price.

Gilbert had come up with his moonlet theory to explain why lunar craters are circular, the old bugaboo to acceptance of meteorite impact on the Moon. In a footnote to his 1927 memorial of Gilbert, in almost a throwaway line, Davis hinted at a way out of the seemingly intractable difficulty: "It has lately been suggested that the circular form of lunar craters might be determined by the explosion of material vaporized by meteoric impact, whatever the direction of a meteor's approach."[16]

* * *

Unbeknownst to Davis, in 1916, a young Estonian astronomer named Ernst Julius Öpik (1893–1985) had come to the same conclusion.[17] Öpik would go on to a brilliant career in astronomy and become honored for his studies of asteroids and comets, as well as for his prediction of craters on Mars and the existence of the giant, distant Oort (or Öpik-Oort) cloud of comets. But it would be forty-seven years after the publication of Öpik's 1916 paper before Ralph Baldwin, in his 1963 book The Measure of the Moon, would cite Öpik's potentially pathbreaking article.[18]

Öpik gave credit for the "meteoric theory" to his mentor, Russian scientist N. A. Morozov, who he said was the first to conclude that impact craters would be circular and bowl-shaped regardless of the angle of an arriving meteorite. As Öpik described Morozov's theory, a "cosmic mass moving with great velocity strikes the surface of our satellite, is smashed to pieces, and buries itself into surface rocks." The impact generates so much heat that "[a] considerable part of the meteoric mass and the surrounding rocks turns into gas which is very resilient due to its high temperature: an explosion happens that scatters around the rock and makes a crater—a bowl-shaped depression with a rim; the meteoric mass generates the central peak."[19] This sequence occurs because of Newtonian physics: the energy of a moving object equals one-half its mass times its velocity squared ($E = \frac{1}{2}mv^2$). Since objects in space travel at hypersonic speeds, even a small one carries enormous energy. Thus, an impacting meteorite would not dig out a crater and bury itself as Gilbert and the Barringers had assumed; it would explode, burst into fragments, and leave a circular crater no matter the angle of impact.

Using the equation, Öpik calculated the size of the meteorite necessary to make a crater of a certain dimension, the same calculation that Moulton would use several years later to determine the size of the Meteor Crater impactor. Öpik found that to excavate the lunar crater Tycho, for example, would have required a meteorite traveling at 30 kilometers per second, weighing 30 million metric tons (30 billion kilograms), and having a diameter of 0.25 kilometer. One could adjust these parameters—a faster meteorite could have been smaller and still have the same energy—but Öpik had shown that an impacting meteorite creates a crater much larger than itself, in the case of Tycho 250 times larger. On Earth, meteorites smaller than around 25 meters burn up in the atmosphere, but larger ones pass through unimpeded, strike intact, explode, and leave a crater whose size is proportional to the size of the impactor.

By originating in an explosion, a meteorite crater comes to resemble the pits left by bursting artillery shells. In a 1919 article, engineer Herbert Ives (1882–1953) compared lunar craters to the explosion pits left by bomb tests during World War I.[20] He evidently did not know of Öpik's paper but came to the same conclusion regarding the Moon's craters.

Ives noted that a bomb equal in power to several hundred pounds of TNT produced a crater that closely resembled medium-sized lunar craters, having "the circular surrounding wall, the central peak, and a few short radiating streaks." Another explosion pit replicated to scale the features of the large lunar crater Copernicus. Ives compared photographs of lunar and bomb craters and, like Öpik, found that no matter the impact angle, the result was a circular crater.

To rebut the claim that the Earth lacks impact craters, Ives pointed out that "Cañon Diablo [Meteor Crater] [is] the most perfect imitation we have of lunar craters." He noted that "upheaval and weathering" could well have removed older craters on Earth or made them unrecognizable.

Bomb explosions—as well as crater analysis—were a sideline for Ives, an unusually versatile and accomplished scientist.[21] Working for AT&T's Bell Telephone Laboratories, he helped develop the facsimile (fax) mode of electronic transmission as well as television transmission systems. In 1938, Ives conducted an experiment that confirmed an aspect of Albert Einstein's special relativity. For his research during World War II on blackout lighting and optical communication, Ives received the Medal of Merit from President Harry Truman.

Another great scientist who became interested in the origin of the Moon's craters was Alfred Wegener (1880–1930). He had proposed the theory of continental drift in 1912, and sixty years later, it became universally accepted as part of the theory of plate tectonics. In 1921, Wegener wrote an article titled "Die Entstehung der Mondkrater" ("The Craters of the Moon."[22] He had read Gilbert's *The Moon's Face* and, like him, conducted his own laboratory experiments. Wegener summed up his conclusion: "The similarity of forms [between lunar craters and terrestrial volcanoes] are [*sic*] totally superficial." To retain the volcanic theory for lunar craters, he said, required an ad hoc, special pleading that amounted to claiming that each similarity between lunar and terrestrial craters confirmed that they had the same origin, while each difference could be ignored as it derived from the inherent dissimilarities between the Moon and the Earth. He ended by saying of lunar and terrestrial craters: "The forms are fundamentally different; therefore, the origins also should be different." He accepted that Meteor Crater was due to impact and wrote, "It is highly improbable that this is the only meteorite crater on Earth."

Another possible candidate for a terrestrial impact crater appeared in a 1926 report of meteoritic iron at a "blow-out" near Odessa, Texas.[23] Barringer's son Reau visited the site and wired his father: "It is a meteor crater, beyond a shadow of a doubt . . . so absurdly like M. C. that it doesn't seem possible."

But if Meteor Crater and the Odessa crater were due to meteorite impact, could they be the only examples on Earth?

6

Cryptic Craters

In 1933, L. J. Spencer, keeper of minerals at the British Museum and a distinguished mineralogist, reviewed the putative meteorite craters worldwide that scientists had so far identified.[1] He included not only Meteor Crater and the Odessa craters (there turned out to be several) but also Wabar in the Rub' al Khali of the Arabian Desert, a group of craters in Estonia on the island of Oesel, another cluster at Henbury in central Australia, the Ashanti crater at Lake Bosumtwi in Ghana filled by a large lake, and a group of small craters at Campo del Cielo in Argentina, the location of the huge meteorite that Chladni had long ago described. Spencer noted the "close analogy" between these and "craters formed by high-explosive shells." He regarded these terrestrial craters as "definite proof" of impact on Earth. But he was oddly noncommittal about lunar craters, saying only that they "are usually thought to be of volcanic origin, but the suggestion has also been made that they were formed by the fall of meteorites."

Gilbert had asked, "Does the earth exhibit impact craters?" and answered, "If not, then erosion and sedimentation have destroyed them." But even an impact crater whose surface features had been entirely removed by erosion might still retain clues to its cosmic origin in its buried and deformed geological structures. One of the first candidates for an ancient, eroded impact structure was the Steinheim Basin in Germany, set inside the larger and better-known Ries Basin. The Steinheim Basin has an uplifted center, a wide depression around its margin, and an absence of volcanic rock. Magnetic surveys suggested the presence of igneous rock a few kilometers beneath the surface, enough for a pair of German geologists in 1905 to ignore the evidence of impact and to declare the Steinheim Basin a "Kryptovulkanische," or cryptovolcanic, structure.[2]

Walter Bucher (1888–1965), who would become one of the most acclaimed American geologists, read their article and took the notion of cryptovolcanism to heart.[3] Born in Akron, Ohio, to Swiss-German parents and raised in Germany, to which the family had returned, he received his PhD from the University of Heidelberg and came back to America to become

Unlocking the Moon's Secrets. James Lawrence Powell, Oxford University Press. © James Lawrence Powell 2023.
DOI: 10.1093/oso/9780197694862.003.0007

a full professor at the University of Cincinnati. In 1925, he published a report for the Geological Survey of Kentucky titled "Geology of the Jeptha Knob," an isolated structure in the state's Bluegrass country.[4] Bucher noted the knob's similarity to the Steinheim Basin and concluded that an "unknown force of deep-seated volcanic origin" had pushed Jeptha Knob upward. He admitted this was a "disturbing" conclusion "in view of the total absence of volcanic activity in Kentucky." Yet, without explaining why, he declared that "the assumption is not unreasonable." In support of his volcanic thesis, Bucher cited a similar uplifted structure at Serpent Mound in Ohio, which also had no associated igneous or volcanic rocks. Again, he cited as evidence the Steinheim Basin, which, he wrote, "has been proven to be connected" with igneous rocks at depth. But geologists had not "proven" any such connection; rather, they had only surmised one. Despite the complete absence of any evidence of volcanism, Bucher had declared each of the three structures—Ries Basin, Jeptha Knob, and Serpent Mound—to be ultimately of volcanic origin.

In 1933, Bucher extended his claim of a cryptovolcanic origin to four other enigmatic geologic structures: Wells Creek Basin in Tennessee, Kentland Dome in Indiana, Decaturville Dome in Missouri, and Upheaval Dome in Utah.[5] Like the first three, each had a "circular outline" and a comparatively small "central uplift surrounded by a ring-shaped depression." They had "no volcanic materials or signs of thermal action" and presented no evidence of "violent action—that is, of an explosive force." Nevertheless, "The evidence of explosive action in the center of these structures is sufficiently convincing to exclude the possibilities of a nonvolcanic origin." This reminds us of the illogical conclusion of the nineteenth-century selenographers, who, despite the total lack of resemblance of lunar craters to terrestrial volcanoes, declared them volcanic anyway. Reasoning from this evidence-free particular to the general, Bucher declared that "All American cryptovolcanic structures represent special phases of the ascent of [basaltic] magmas into the central plateau region of the United States." Thus, we see how his assumption regarding tiny Jeptha Knob was transmogrified into a fact applicable everywhere on Earth.

Until near the very end of his life, Bucher never accepted the presence of meteorite impact craters on Earth and continued to embrace the cryptovolcanic theory. A year before he died, however, after a 1964 field trip to Meteor Crater with Eugene Shoemaker, Bucher acknowledged that it might be a terrestrial impact crater after all, but as they left, he added, "Ah, but the Ries—that's still different!"[6] But by this time, Shoemaker had already shown that the Ries Basin is an impact crater. It would soon be used to train the Apollo 14 astronauts to recognize impact products.

* * *

Bucher's claims did not convince all geologists. In a series of prescient articles during the late 1930s, two geologists from Southern Methodist University, John D. Boon (1874–1952) and Claude Albritton (1913–1988), concluded that meteorites larger than 100 feet in diameter would pass through the atmosphere unimpeded and "must *explode* when they strike the earth [italics in original]," the same conclusion that Öpik had reached, though they did not know of his work.[7] The two argued that long after erosion had removed the surface evidence of impact, subterranean structures might reveal that they had originated in an explosion. They included a diagram to show what they had in mind (see Figure 6.1).[8] The different cross sections—AA to DD—show imaginary descending levels of erosion. AA, at the surface, could represent Meteor Crater, but as a fresh crater erodes through millions of years, the surface level drops continuously, leaving only a formerly buried, distorted structure to reveal its meteoritic origin. Below DD, Boon and Albritton wrote, "the scar would be obliterated." If you focused on section CC, say, and

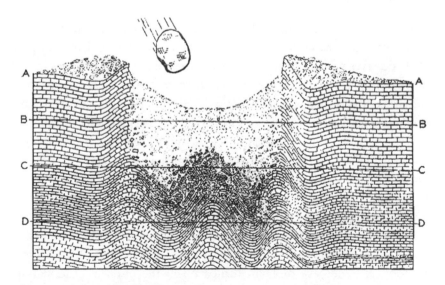

Figure 6.1. Erosional levels in an impact crater. (Boon and Albritton, "Meteorite Scars.")

imagined everything above it gone, it would look remarkably like the cross section of one of Bucher's cryptovolcanic structures.

As evidence for his claim that subterranean volcanism created all such structures, Bucher noted that the six he described in his 1933 paper lay on the sides of giant structural rock domes hundreds of kilometers wide. He interpreted this to mean that the structures were not situated at random, as meteorite impact craters would have to be, but rather that the underlying geology had determined their location. Boon and Albritton countered that such domes and other large structural features are common at the Earth's surface, so there was a good chance that a meteorite arriving at random would land on one.

Like most geologists of his era and training, Bucher unequivocally accepted uniformitarianism, the view that geological change operates gradually and in a uniform way. Meteorite impact could almost define a catastrophic event, the very antithesis of uniformitarianism. Moreover, geologists preferred not to appeal to the deus ex machina of extraterrestrial events. In 1963, Bucher would warn his colleagues not to "look to the sky to solve our problems."[9] He went to his grave opposing not only meteorite impact but also the theory of continental drift, another catastrophic process that violated uniformitarianism.

* * *

Robert Dietz (1914–1995) was Bucher's scientific opposite, a creative iconoclast who accepted both meteorite impact and continental drift.[10] Indeed, Dietz deserves partial credit for the hypothesis that the ocean floors spread outward from the midocean ridges and may carry the continents piggyback, the core of the theory of plate tectonics.[11] He was one of the most original thinkers in geology during the middle decades of the twentieth century.

Dietz became interested in astronomy at an early age and his fascination stayed with him for life. In graduate school at the University of Illinois, he proposed for his PhD thesis topic a study of craters on the Moon and the Earth, but his professors turned him down, urging him to focus instead on marine geology, which he did with great success. But a serendipitous event would lead him back to his long-standing curiosity about impact craters. In August 1941, Dietz received his military draft letter and applied for flight school.[12] A minor eye problem turned up in the physical examination, leading to his rejection. In response, he drove 320 kilometers to take the exam again, and this time he passed. A very persuasive fellow, Dietz may well have talked his way past his eye problems. As luck would have it, during his 1943 flight training, Dietz found himself near Kentland, Indiana, the location of one of Bucher's cryptovolcanic structures. The imaginative Dietz, his interest in lunar and terrestrial craters rekindled, wondered whether the Kentland structure might be "an asteroidal impact scar." To test the hypothesis, he visited the Kentland quarry, where he found peculiar structures called shatter cones. As shown in Figure 6.2, they resemble a horse's tail or badminton shuttlecock but in solid rock. When "the vertically dipping strata were rotated [in his imagination] to an assumed original horizontal position," Dietz wrote, "the cones pointed upward suggesting that the fracturing impulse came from above and hence was cosmic rather than volcanic"[13] in origin. This would later lead him to propose that the Kentland structure was an eroded meteorite impact scar and had nothing to do with volcanism. Scientists would come to accept shatter cones as distinctive indicators of meteorite impact.

In 1946, Dietz wrote what he would later say was "[t]he first paper in a geological journal to suggest neither a volcanic nor a tectonic origin [for lunar craters] since that of G. K. Gilbert,"[14] half a century earlier. Evidently, Dietz did not know of the articles by Öpik, Ives, and Wegener or of one from 1942

Figure 6.2. Shatter cone from the Steinheim Basin, Germany. (Wikimedia Commons.)

by Ralph B. Baldwin (1912–2010) titled "The Meteoritic Origin of Lunar Craters."[15] In those days before online journals, it was entirely possible for a scientist to miss articles.

Dietz did note that crater circularity is "the result of being formed by explosion" and that "a relatively small meteorite can produce an enormous crater in the lunar environment."[16] The dark regions of the Moon's surface, Dietz said, "are probably extensive lava plains generated by the impact of bodies of asteroidal dimensions and were formed relatively late in lunar history." The articles by Gilbert, Öpik, Wegener, and Dietz had something other than their subject in common, however: no one paid them any attention.

* * *

That was not true of Baldwin's pathbreaking 1949 book, *The Face of the Moon*.[17] One reason the book caught the attention of scientists was that Baldwin included data that could be used to test whether explosion had created lunar craters. He assembled information on the depth, diameter, and rim height of 329 lunar craters as measured by telescopic observations. He noted that the diameters of the largest lunar craters exceeded those of the smaller ones by 150 times, whereas the bomb craters were 13 times smaller than the smallest lunar crater. To show such a wide range on one chart, he used the "log-log" plot, the logarithm of depth plotted against the logarithm of diameter as shown in Figure 6.3. Solid squares represent terrestrial impact craters and one nuclear test site (Sedan). "Unfortunately, the [bomb craters and lunar craters] do not overlap," Baldwin noted. But "Nature has stepped into the breach and obligingly furnished four meteoritic craters whose dimensions have been carefully measured." He included the four—Barringer (Meteor Crater), Henbury, and two Odessa craters—on his chart. In a replotting of Baldwin's original chart, C. S. Beals and his colleagues added four Canadian impact craters: Brent, Deep Bay, Holleford, and New Quebec.[18] I have added to their plot the position of the crater from the 100-kiloton Sedan underground nuclear test in 1962.

"The only reasonable interpretation," Baldwin concluded, "is that the craters of the Moon, vast and small, form a continuous sequence of explosion pits, each having been dug by a single blast. No available source of energy is known other than that carried by meteorites." He summed up: "To claim that the Moon's craters are volcanic is tantamount to postulating an entirely

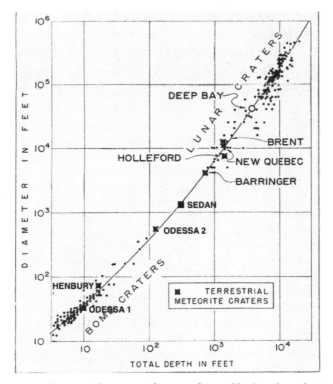

Figure 6.3. Depth versus diameter of craters formed by bomb explosions, nuclear tests, and meteorite impact. (Baldwin, *The Face of the Moon*; modified).

new, entirely hypothetical mode of origin and to fly in the face of the fact that a known process is completely able to explain the vast majority of observed lunar features." Baldwin waxed eloquent in wondering about impact craters on Earth: "How many other gigantic meteorites have disappeared into the watery wastes? What myriads of meteoritic craters lie unseen on the surface or for ages hidden in stony crypts?"

In a 2003 interview with Baldwin, American geologist and author Ursula Marvin told him that *The Face of the Moon* had "launched research on the Moon as a modern branch of science" and had become "that rarest of things, a book that made a difference."[19] Indeed, the book attracted more than a few readers whose lives it would change and who in turn would change our understanding of both the Moon and the Earth.

One was Harold Urey (1893–1981), winner of the 1934 Nobel Prize in Chemistry for his discovery of deuterium, the second isotope of hydrogen. He would go on to become one of the key scientists on the Manhattan Project and in NASA's Apollo program. Another was Shoemaker, who had solved the mystery of Meteor Crater.

7

To the Moon

Eugene Shoemaker's interest in the Moon, like that of Robert Dietz, began early.[1] He remembered that on his twentieth birthday, he had looked up at the Moon and thought:

> I want to go there! I want to be one of the first people on the Moon. Why will we go to the Moon? To explore it, of course! And who is the best person to do that? A geologist, of course! I took the first fork that went to the Moon that morning.

Shoemaker grew up in Los Angeles, where visits to the famous La Brea Tar Pits Museum sparked the lad's interest in science. He earned a bachelor's degree from the California Institute of Technology and after graduation took a job with the US Geological Survey, which allowed him to complete a PhD at Princeton. He began his work with the Survey at the volcanic Hopi Buttes in northern Arizona, searching for the uranium deposits that had suddenly become important in the Cold War. The buttes were not far from Meteor Crater, which Shoemaker visited and which caught his interest. The Barringer family still owned the crater, and Shoemaker was aware of their antipathy toward the Survey and its obstinate denial of an impact origin for Meteor Crater. The story goes that he won over the Barringers by having a friend of the family vouch for him.

Shoemaker epitomized the modern geologist, skilled at field mapping but also adept with complex mathematical equations. In his classic 1959 paper "Impact Mechanics at Meteor Crater, Arizona,"[2] he noted that most previous work (like Moulton's) had focused on the theoretical size of the putative meteorite. Instead, he would focus on the geology of Meteor Crater and its structural similarity with a nuclear explosion crater.

Shoemaker made an exhaustive study of the rock types in and around Meteor Crater. He came to know each geological formation so well that he could identify it just from small rock fragments in the rim. This allowed him to become the first to discover that the "bedrock stratigraphy"—the order

Unlocking the Moon's Secrets. James Lawrence Powell, Oxford University Press. © James Lawrence Powell 2023.
DOI: 10.1093/oso/9780197694862.003.0008

by age of the rock layers— "is preserved, inverted, in the debris walls [in the rim]." In other words, the strata around the edge of Meteor Crater stack bottom-side up, with the older beds on top of younger ones. Some enormous force had pushed the bedrock up and out and flipped it over to turn the section upside down. As Shoemaker described it, "Rocks now represented by the debris of the rim have been peeled back from the area of the crater somewhat like the petals of a flower blossoming."

Having been detailed from the USGS to the Atomic Energy Commission, Shoemaker knew of the results of underground nuclear tests, enabling him to recognize their similarity to Meteor Crater. At the Teapot Ess Crater, produced in 1955 by a 1.2-kiloton device detonated at a depth of 20 meters, Shoemaker saw that "[b]eds exposed in the rim are peeled back just as the bedrock is peeled back at Meteor Crater."

Shoemaker turned up other evidence of impact at Meteor Crater, including meteorite particles encased in once-molten quartz, which had also been found at the Wabar and Henbury craters, suggesting high-pressure shock. A different and perhaps even more convincing kind of evidence came from the discovery at Meteor Crater of two rare varieties of silicon dioxide, coesite and stishovite, which form only under extreme shock pressure. Scientists had synthesized coesite in the laboratory but had never found it in nature until Shoemaker sent a specimen from Meteor Crater to a colleague at the USGS, who found it to contain both minerals. Thus, by 1960, Shoemaker had established the meteoritic origin of Meteor Crater and scientists had identified a dozen or so other likely terrestrial impact structures. As for lunar craters, they remained comfortingly out of reach, allowing plenty of elbow room for the proponents of lunar volcanism to persist in their beliefs.

* * *

On May 16–19, 1964, the New York Academy of Sciences held a conference titled "The Geological Problems in Lunar Research." Scientists use the word "problem" to stand for "unanswered questions," those that drive further research. In this case, the main problem at issue was the now-centuries-old question of the origin of the Moon's craters. As we will see, this may have been one of the most ill-timed scientific conferences ever. One of the organizers, Jack Green, had been Bucher's graduate student, and where the teacher led the student followed. "Early in this century," Green quipped, "astronomers apparently rebelled against the anagram of their name—moonstarers—by deliberately discouraging graduate students from investigating the moon."[3]

Green deplored "dogma" in lunar studies but then stated, "Meteoritic impact is considered to be a trivial process in affecting both the genesis and development of almost all major lunar surface structures." Another conference paper, "Proof of the Volcanic Origin of Most Lunar Craters and of Tectonic Maria," concluded with this dogmatic statement: "Very few phenomena, if any, and at best only the very smallest among them, can be explained by the impact theory."[4] Thus, even as late as 1964, with spacecraft on their way to the Moon, most lunar scientists continued to endorse the view stretching back to Hooke: volcanism had created the Moon's craters.

Bucher titled his conference paper "The Largest So-Called Meteorite Scars in Three Continents as Demonstrably Tied to Major Terrestrial Structures." It would be his last scientific publication. He again argued that since cryptovolcanic structures are not randomly distributed, they cannot be due to extraterrestrial events. This conclusion required him to reject all other evidence of impact. "The presence of shatter cones or of coesite in structures developed by powerful impact cannot be accepted as sufficient evidence for impact from above," he said. "All three structures [Meteor Crater, the Ries Basin, and Vredefort Dome, a giant, ancient ringed structure in South Africa where Dietz had found shatter cones] are so closely associated with deep-seated volcanic phenomena, that appeal to chance in all three cases seems quite inadmissible."[5] But this "close association" with volcanism appears to have been largely in Bucher's mind. His illogical rejection of impact on the Earth remained unchanged from when he had studied little Jeptha Knob almost forty years before.

* * *

The meeting took place during NASA's Ranger program, designed to carry spacecraft to the Moon to take close-up photographs and telemeter them back to Earth, just before the vessel crashed and destroyed itself. NASA began the program in 1961, only months after President John F. Kennedy had told Congress on May 25, "I believe that this nation should commit itself to achieving the goal, before this decade is out, of landing a man on the moon and returning him safely to the earth."[6] To meet that goal, mission planners had to know whether the surface of the Moon would support the footpad of a lunar lander, in which case it would also be strong enough to bear the weight of an astronaut. US officials had justified the cost of the space program not merely as necessary to beat the Russians to the Moon but also for the advances in science that they believed would result. Merely viewing the

Moon from an orbiting spacecraft would not accomplish that goal; astronauts had to walk on its surface, collect samples, and bring them back to the Earth. But some scientists, including the iconoclastic astrophysicist Thomas Gold, had concluded that a thick and dangerous layer of dust coated the Moon.

Like much of the early US space program, Ranger got off to a disappointing start. The first five attempts failed to get off the launch pad, missed the Moon, or impacted on the farside. On January 30, 1964, Ranger 6 was on its way to success, when its cameras failed. Then came Ranger 7, which on July 28, 1964, had a perfect launch into the trajectory that would take it to the Moon. Three days later, the craft came within 900 kilometers of the lunar surface, the distance at which its cameras were to turn on. Soon the mission announcer reported, "We have video." As Ranger returned the first photograph of the Moon taken by a US spacecraft (see Figure 7.1), Jet Propulsion Laboratory engineers leaped to their feet in joy and amazement.[7] Ranger 7

Figure 7.1. First close-up photograph of the Moon from a US spacecraft. The large crater at center right is the 108-kilometer Alphonsus, with its bright central peak. (NASA.)

struck near its target in the northwest section of Mare Nubium, the Sea of Clouds. The plain in which it landed was soon named Mare Cognitum, "The Sea that has Become Known." It would become the landing site of Apollo 12.

Ranger 7's six cameras sent back 4,300 images to joyful scientists and engineers, showing features two thousand times smaller than those that could be seen through earthbound telescopes. We rightly recognize Galileo and his telescope as having launched a new era in astronomy. The Ranger 7 images can well stand as the emblem of the age of space exploration. The final Ranger 7 photos (see Figure 7.2) showed a Moon carpeted with craters on every conceivable scale, down to about 50 centimeters, the limit of resolution. This proved Wegener right yet again: "The face of the Moon is covered with craters to such a degree that it seems doubtful to be able to find a single point which at least once has not been a part of the floor of a crater."[8]

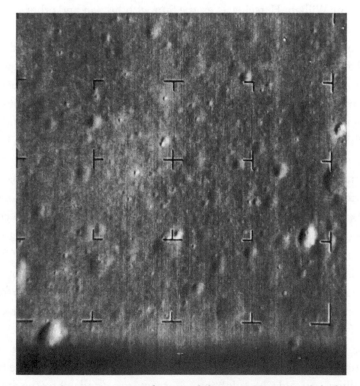

Figure 7.2. Final Ranger 7 image from 6.5 kilometers, 2.5 seconds before impact. Small craters down to the limit of resolution pepper the flat floor of Mare Nubium. (NASA.)

Working double time, on August 28, 1964, a team of scientists produced an interim report on their analysis of the Ranger 7 photos.[9] They found that regardless of size, most lunar craters result from meteorite impact, thus rendering the report from the New York Academy of Sciences conference mostly obsolete before the ink had dried.

* * *

Before we move on to the origin of the Moon itself, let us look back at the main themes of this first half of the book. Belief that the Moon harbors life, even life much like that on Earth, had persisted from the time of Kepler in the early 1600s to at least a 1921 report by astronomer William Henry Pickering, director of the Harvard College Observatory, who said he had observed vegetation on the Moon and believed that it would support two crops a day, earning him a lengthy favorable article in the *New York Times*.[10] Along the way, many eminent scientists populated the Moon, the planets, and, in the case of Sir William Herschel, even the Sun with human and other advanced life forms.

From Hooke in the 1600s until the mid-1960s, scientists continued to profess that the forces that had created lunar craters had come from below, from volcanism, even though lunar craters differed in nearly every respect from terrestrial ones. The proponents of volcanism brushed aside each difference as due to the differing physical conditions on the two bodies. To dislodge the belief that lunar craters are volcanic required more than observation by telescope; spacecraft had to go to the Moon and send back close-up photographs showing a surface blanketed with craters.

As for terrestrial impact craters, scientists later determined that each of the six structures that Bucher had called cryptovolcanic had some combination of impact indicators, such as coesite, elevated iridium, shatter cones, and quartz shocked at hypersonic pressures. By the end of the 1960s, few scientists doubted that meteorite impact had created nearly all lunar craters and that many impact structures exist on Earth, though none was as fresh as Meteor Crater, believed to be only about fifty thousand years old. In recognition of the growing rejection of "cryptovolcanic" structures, scientists began to use the term "cryptoexplosion," but it did not survive for long. Today scientists have confirmed 190 impact craters on Earth, including the buried 66-million-year-old Chicxulub impact structure blasted out by the dinosaur killer. To imagine how many more impact craters must have once existed on our planet before erosion and plate tectonics destroyed them, we need only look at photographs of the pocked faces of the Moon and Mercury.

The great physicist Richard Feynman said, "The first principle is that you must not fool yourself—and you are the easiest person to fool."[11] The history that we have reviewed shows how generations of scientists fooled themselves into believing in extraterrestrial life, that lunar craters were volcanic, and that there were no impact craters on Earth. Few who held these views ever publicly changed their minds, instead carrying their beliefs to the grave. Scientists can fool themselves, as can anyone, but science is self-correcting and in the long run cannot be fooled. Nothing shows this better than the subject we take up next: how scientists continually tested, improved, and, when necessary, rejected a succession of theories for the Moon's origin, until they finally discovered one that worked.

PART II
THE ORIGIN OF THE MOON

8

The Rise and Fall of the
Nebular Hypothesis

In this second part of the book, we turn to our most fundamental question: where did the Moon come from? Scientists could not answer that question until they had established that meteorite impact, not volcanism, had
created lunar craters, which, as we have seen, did not occur until the 1960s. But
speculation about the origin of the Moon and the planets began much earlier,
soon after Newton's discovery of gravity. In 1665, while Newton (1643–1727)
was a student at Trinity College, Cambridge, a pandemic known as the Great
Plague of London led him to shelter at his family farm. Surely, no one has ever
put enforced solitude to better use than he did. During what some have called
his annus mirabilis, or "year of wonders," he worked on optics, developed
calculus, and pondered the force that maintains the Moon in its orbit, alleg-
edly inspired by an apple falling from a tree. In his *Principia Mathematica*,
published in 1687, Newton's early thinking bore fruit in the universal law of
gravitation: every particle in the universe attracts every other particle with a
force proportional to the product of their masses divided by the square of the
distance separating them. Newton used calculus to show that the law of gravitation explained the motion of the planets around the Sun and of their moons
around them, giving rise to a new field of science called celestial mechanics.

Newton explained why the Moon orbits the Earth, rather than shooting
off into space or, oppositely, colliding with our planet. The Moon appears
to hang in space due to the balance of two fundamental forces. One is momentum: the mass times the velocity of a moving object. Were it the only
force acting, the Moon would naturally continue in a straight line past the
Earth and fly off into space. But gravity pulls on the Moon until it just balances momentum, locking it into orbit. The two bodies also pull tides in each
other, and to conserve energy, a principle that Newton also discovered, the
Moon recedes slightly and the Earth slows slightly in concert.

* * *

Unlocking the Moon's Secrets. James Lawrence Powell, Oxford University Press. © James Lawrence Powell 2023.
DOI: 10.1093/oso/9780197694862.003.0009

In *Exposition for the System of the World*, published in 1796, Pierre-Simon Laplace (1749–1827) used Newtonian gravity to explain the origin of the solar system.[1] He listed five facts (as known at the time) that any theory of origin had to explain:[2]

1. The planets move around the Sun in the same direction and in orbits that lie almost in the plane of the path of the Sun in the sky, which, as noted earlier, astronomers call the ecliptic.
2. The satellites move in that same direction around their primary bodies, in a plane close to the ecliptic.
3. The planets and their satellites rotate on their axes in the same direction as their revolution around the Sun, also in nearly the same plane. Astronomers call this prograde motion.
4. The orbits of the planets and moons are nearly circular.
5. Comets come from outside the solar system and travel in elliptical orbits that incline at random to the ecliptic.

Laplace presented a model of how the solar system might have evolved to account for these observations. He postulated that the Sun's atmosphere had once been a hot, rotating cloud of gas covering the entire region of what is now the solar system. Gravitational attraction converted it into a flat disk, and as it cooled and contracted, the law of conservation of momentum (implied by Newton's laws of motion) caused it to rotate faster, like the proverbial ice skater who draws in her arms as she spins. As the cloud spun, centrifugal force at the outer edge of the disk eventually overcame the pull of gravity toward the center. At that point, a ring of gas separated while the remainder of the mass continued to contract and spin inside the ring. According to Laplace, the process repeated itself several times, each ejection producing a new ring that subsequently condensed into a planet. The planets then recapitulated the process, producing rings that collapsed to become their moons. Gravity held the entire clockwork arrangement in place. Astronomers had long observed fuzzy regions in space that they called nebulae (Latin for "cloud" or "fog"), which they thought represented the stages in cosmic evolution that Laplace described. Thus, his model became known as the nebular hypothesis.

The son of a middle-class family in Normandy, Laplace showed an unusual ability in mathematics from an early age, soon writing articles that in 1773, when he was only twenty-four, led to his election to the Académie Royale des Sciences. From there he was awarded a professorship at the École Militaire, where the young Napoleon was one of his students. Laplace went on to make several breathtaking advances in science. As part of his hypothesis, he refined how Newton's law of gravity explained the observed motions of all the moons and planets. This meant that God no longer needed to intervene to hold the planets in their orbits, leading Laplace "to the final removal of divine influence from the present operations of nature."[3] He also developed the mathematical analysis of probability and was likely the first to conceive of what came to be called a "black hole," a star so dense that its gravity blocks any light from escaping.

It is no surprise that, like most theories in science, the nebular hypothesis had precedents. In 1775, the German philosopher Immanuel Kant proposed that the solar system came into existence first as a large cloud of gas, which gravity had caused to clump into denser regions and eventually into particles and larger objects; these then accreted further to become the planets and moons. Kant was also the first to point out that the Moon's gravity would slow the Earth's rotation until eventually the Moon kept the same face turned to the Earth, the problem identified in the Introduction. But because Laplace developed the nebular hypothesis much more fully than Kant, he is given priority.

Another predecessor and the only one whom Laplace cited in his *Exposition* was Georges-Louis Leclerc, Comte de Buffon (1707–1788), who wrote a thirty-five-volume treatise on natural history. Buffon postulated that collision with a comet had caused the Sun to eject a long streak of matter that eventually became the planets. While still molten and spinning rapidly on their axes, the primordial planets had in turn flung off smaller globules that became their moons. The hypothesis inspired Buffon to build iron balls of different sizes, heat them up, and time how long each took to cool. Extrapolating from his laboratory balls to one the size of the Earth, he found that our planet would have been molten 3 million years ago. This he equated with the age of the Earth's formation. But the cautious Buffon lowered his published result to seventy-five thousand years, still too much for the theological faculty of the Sorbonne, who required him to reduce the age still further. Buffon's theory could account for the first of Laplace's five facts listed above but not the other four.

* * *

The nebular hypothesis dominated nineteenth-century science and became accepted as a given, the starting point for all other theories. It also foreshadowed what we might call the evolutionary view of the solar system to go along with the biological evolution that would come decades later with Darwinism.

Laplace's countryman Édouard Roche (1820–1883) later calculated the gravitational attraction between a primary planet and its smaller satellite.[4] He found that the smaller body cannot approach closer than 2.44 times the radius of the primary body before the gravitational pull of its larger neighbor overcomes the gravity that holds the smaller body together and tears it into fragments. This distance, known as the Roche limit, remains an important element of modern theories of the origin of the Moon. Like Öpik on the meteoric origins of lunar craters and Gregor Mendel with his pea experiments, Roche published in an obscure journal, postponing awareness of his work by a wider scientific audience.

In 1873, Roche worked out mathematically how larger bodies might cast off rings like those around Saturn, which would then condense into satellites, as the nebular hypothesis proposed. His analysis explained why Mars, Venus, and Mercury had no moons (as was then believed) and why no planet lay between Mars and Jupiter, where we should expect one by "Bode's law," formulated in 1772, according to which each planet lay about twice as far from the Sun as the one closer in. Roche showed that the absence derived from the enormous mass and gravitational power of Jupiter, which broke apart and swallowed up any incipient planet that came within reach of its powerful gravity.

The history of the nebular hypothesis provides a classic example of how scientists can continue to accept a theory despite anomalies but only up to a point. In 1877, Mars lay in "opposition," that is, on the opposite side of the Sun from Earth, making it possible for astronomers to see that the Red Planet has two tiny moons, Phobos and Deimos. But the motion of Phobos has a peculiarity: it completes one revolution of Mars every 7 hours, 38 minutes, while Mars takes 24 hours, 37 minutes to rotate on its axis, roughly three times as long. Yet according to the nebular hypothesis, a planet should rotate faster than any ring or satellite that it has flung off. Phobos showed just the opposite, a discrepancy that might well have led scientists to abandon the nebular hypothesis. Instead, a typical reaction was that of American astronomer Daniel Kirkwood in 1878: "The amount of evidence in favour of the nebular hypothesis has been generally regarded as giving it a degree of probability little short of demonstration [proven]."[5] To overthrow the hypothesis, he said, would take "nothing less than facts absolutely incompatible with its essential conditions," and in his mind, the speed of Phobos's revolution did not rise to that level. It may nevertheless have had a psychological effect, for a year later, Kirkwood did identify what he thought were fatal flaws in the nebular hypothesis. This tipped the scale, causing him to conclude that either the "celebrated hypothesis must yet be abandoned, or its principal features must be essentially modified." He thought no modifications could save it, however.[6] The nebular hypothesis continued to falter, and by the end of the nineteenth century, it had to make way for new ideas that could explain more of the evidence.

* * *

With the nebular hypothesis off the table, during the late nineteenth and early twentieth centuries, scientists developed three independent theories for the origin of the Moon: fission, co-accretion, and capture. Through the decades right up to the present, one theory would rise in favor and seem to be vindicated, only to fail to explain some piece of evidence and have another one supplant it. That theory in turn would weaken, and the process would repeat. Science historian Stephen Brush uses this history to make an important point: "The approach in science textbooks and popular articles is to start from the present theory, assuming it to be correct, and ask how we got there." But this view "is not very satisfactory if it has to be rewritten every time the 'correct answer' changes."[7]

According to the nebular hypothesis, the centrifugal force of a rotating, molten planet caused it to spin off material that coagulated to become its satellite. George Howard Darwin (1845–1902), son of Charles, first developed the fission theory mathematically. He studied at Cambridge, won second prize in the annual student math contest (as had the great physicist William Thomson, Lord Kelvin), and went on to become the Plumian Professor of Astronomy at the university. He assumed the presidency of the Royal Astronomical Society in 1899 and in 1905 was elected the president of the British Association. King Edward VII made him Knight Commander of the Bath.

In 1879, George Darwin proposed that the proto-Earth had spun much more rapidly than it does today, so that the day was only a few hours long.[8] The centrifugal force due to the rapid spin allowed the tidal bulge pulled on the proto-Earth by the gravity of the massive Sun to break off and become the Moon. After that, the tides that the Moon pulls in the Earth's oceans slowed the Earth's rotation, as Kant had suggested and as conservation of energy requires. But because of another discovery of Newton's—angular momentum—the slowing meant that the Moon gradually receded from the Earth. Here is how it worked. In Newton's mechanics, the larger the mass and the higher the velocity of a moving object, the greater the momentum. It is possible to show from Newton's laws that momentum, like energy, is a "conserved quantity": it is neither lost nor gained. One familiar demonstration occurs when a cue ball smacks into the racked balls on a pool table and stops dead, transferring its momentum to them. Objects traveling in a straight line have momentum, but so do those that orbit another body, and we call this angular momentum. The Earth has an orbital angular momentum derived from its revolution around the Sun and a spin angular momentum from its rotation around its axis. The Earth's total angular momentum is the sum of the two. The Moon has angular momentum from its revolution around the Earth and its axial spin. As tidal forces from the Moon pull on the Earth and slow its rotation, to conserve the total angular momentum of the pair, the Moon must recede from the Earth.

Darwin's clever idea was, in effect, to imagine this process running backward in time. He calculated the Earth-Moon distance as it gradually shrank until the two bodies were only about three Earth radii apart. According to Darwin's mathematics, this proximity took place 56 million years ago, which therefore must be roughly equivalent to the age of the Earth and the Moon. This figure came acceptably close to the 100-million-year-estimate of the age of the Earth by Kelvin, so that each calculation seemed to independently confirm the other. Darwin gave little credence to his model for the origin of the Moon, writing, "There is nothing to tell us whether this theory affords the true explanation of the birth of the moon, and I say that it is only a wild explanation, incapable of verification."[9] Yet, as we will see, the fission theory remains an element of the latest models of the origin of the Moon.

Reverend Osmond Fisher (1817–1914), an English geologist, developed a variation of the fission theory in which the Earth's thin, solid crust rested on a layer of molten rock. In an 1882 article in *Nature*, he proposed that the Moon's violent departure from the Earth had left a giant scar that became the Pacific Ocean basin.[10] Then the "cooled granitic crust" moved over the subterranean liquid layer to fill in the cavity, fragmenting the crust into the continents. Fisher wrote, "This would make the Atlantic a great rent, and explain the rude parallelism which exists between the contours of America and the Old World." In other words, Fisher grasped the basic concept of continental drift (though not its cause) before Wegener, who cited Fisher's work. But Fisher was wrong about the Moon's origin and the formation of the Pacific Ocean basin and did not develop his idea of how the continents formed, giving Wegener rightful credit for continental drift.

Roche invented what became known as the co-accretion or binary-planet theory. After the Sun threw off a bulge of nebular material that became the Earth, in his view, the gas-rich early Earth spun off material that, because it lay inside the Roche limit, broke into fragments and formed a ring like those of Saturn, from which the Moon later coalesced. Thus, the Moon and the Earth formed from the same material at the same distance from the Sun. Co-accretion never became compelling enough to dislodge fission, and then, early in the twentieth century, a third theory came to threaten both. It appeared in 1910, when a controversial astronomer named Thomas Jefferson Jackson See (1866–1962) published at his own expense a magnificent, equation-filled, beautifully illustrated, quarto-sized book of 735 pages: *Researches on the Evolution of the Stellar Systems: The Capture Theory of Cosmical Evolution*.[11] Long before others, See accepted meteorite impact as the cause of crater formation on the Moon and a natural outcome of what became known as the capture theory of lunar origin.

See posited that rather than the Sun throwing off the planets, as the nebular hypothesis envisioned, each planet had approached from some distant part of the solar nebula, close enough for the Sun's gravity to capture them into orbit. Similarly, the smaller moons had also circled the Sun originally, but the gravitational pull of the several planets had reeled them in. Thus, our Moon had grown up somewhere else in the solar system but had been captured by the Earth's gravity.

For years, See had doubted the nebular hypothesis but had no substitute; then he had hit on the idea that a resistant gas had filled the primordial nebula and had affected the motion of the planets and moons. He introduced his capture theory in an article in *Popular Astronomy* in 1909, the year before *Researches* appeared."[12] See began by paying obligatory tribute to Laplace and the nebular hypothesis, then quickly shifted gears to pronounce it "easily proved to be untenable and altogether devoid of foundation." He cited several facts that contradicted the nebular hypothesis, ones that he said were the result of the resistant interplanetary gas that he had imagined. As one example, See claimed that the planets had originally traveled in highly eccentric (noncircular) orbits, but the nebular gas had obstructed them and brought their orbits close to circularity. The capture theory did away not only with the nebular hypothesis, See claimed, but also with the fission theory, as

"The Theory of the Resisting Medium [has] effects exactly opposite to those due to tidal friction."

See concluded his book in the grandiose fashion typical of him: "The old theories rest on false premises which must be permanently given up. The capture of the Moon by the Earth may therefore be regarded as a demonstrated fact, which seems destined to become an accepted item of scientific philosophy." Few would go that far, but See did get the capture theory on the list of plausible theories for the Moon's origin, where it has remained in one form or another up to the present day.

* * *

By the end of the nineteenth century, a salvo of slings and arrows had brought the nebular hypothesis to its knees. Noted scientists including James Clerk Maxwell had abandoned it; then, in 1900, came the coup de grâce in an article by Moulton, the astronomer who had calculated that the Meteor Crater impactor was too small to mine and in any case would have broken into widely dispersed fragments.[13] Moulton had graduated from Michigan's Albion College and gone on to graduate school at the University of Chicago, where he encountered See (whom he would later formally charge with plagiarism). At the university, geologist Thomas Crowder Chamberlin (1843–1928) spotted Moulton's talent and later invited him to join a project to study the origin of the Earth. Moulton began his 1900 article by paying homage to the nebular hypothesis, "One of the boldest and most attractive speculations ever offered in any science . . . accepted almost without question by the highest authorities, as Helmholtz, Kelvin, Newcomb, and [George] Darwin." Indeed, Moulton gave credit to the hypothesis as "the first explicit formulation of the theory of evolution," by which he included biological evolution.

Moulton went on to question whether Laplace's nebular hypothesis was "compatible with the fundamental laws of dynamics," of which the conservation of angular momentum is one of the most important. A fundamental fact of cosmology well known in Moulton's day is that although the Sun holds 99.9 percent of the mass of the solar system, it has less than 4 percent of the total angular momentum, the opposite of what we would expect had the Sun spun fast enough to fling off blobs that became planets. After a detailed analysis, Moulton concluded that the "numerical discrepancies [in angular momentum] are so great that it seems to render the nebular hypothesis absolutely untenable."

Moulton came to the defense of Chamberlin, who had the nerve to claim that the Earth is much older than the 100 million years calculated by Kelvin, the ruling authority. Following the nebular hypothesis, Kelvin had envisioned the Earth starting out molten and cooling with no other source of heat than the original. Moulton wrote that without the nebular hypothesis, Kelvin's method had lost its starting assumption, opening the way to Chamberlin's more expansive timescale and providing geologists with the many hundreds of millions of years they believed were required to explain Earth history.

Chamberlin had not only successfully taken on Kelvin, but in a joint effort, he and Moulton had invented a worthy replacement for the nebular hypothesis. The Chamberlin-Moulton "planetesimal hypothesis" began with a star that passed near enough to the Sun to pull out molten bulges that formed two long, curving strands of solar material, like those that astronomers were beginning to see in the newly discovered spiral nebulae. Much of this material fell back into the Sun, but enough remained to cool and coagulate into many minute objects, or "planetesimals." According to the theory, these had collided with each other and over time had accreted to construct the planets and moons, with the leftover bits becoming the icy comets and rocky asteroids.

The planetesimal hypothesis had two decades of success and elements of it remain in the latest theory for the origin of the Moon. But in the mid-1920s, pioneering astronomer Edwin Hubble (1859–1963) observed that the nebulae are not proto–solar systems as Chamberlin and Moulton had believed but galaxies like our Milky Way, only much larger and farther away.

For three decades following Hubble's discovery, the three classic theories of the origin of the Moon waxed and waned in favor as scientists applied new methods to the old problem. They would have gladly applied new data, but there were hardly any—most of what scientists could discover about the Moon from 435,000 kilometers away they had long known. Moulton's calculations had shown that the nebular hypothesis could not explain the angular momentum of the Earth-Moon system, but neither could any of the three theories of the Moon's origin. Even today, angular momentum remains a formidable obstacle to a successful theory of the Moon's origin.

* * *

One property that had to be explained is the Moon's lower density: 3.3 g/cm³ compared with 5.5 for the Earth. This favored the capture theory, as a proto-Moon coming from somewhere else in the solar system might well have arrived with a different density from that of the Earth. However, calculations showed that capture was highly improbable. A wayward voyager would have had to travel at just the right speed and on just the right trajectory, or it would either miss Earth entirely or crash into it. This was not impossible, but the odds were against it.

Fission could explain the Moon's lower density since it might have been flung from the outer, less dense regions of the proto-Earth. Yet fission would have left the early Earth-Moon system with four times the angular momentum that the two combined have today, with no obvious way of getting rid of this conserved quantity.

The Moon's density appeared to pose a near-fatal contradiction to the co-accretion theory, since bodies that arose in the same part of the primordial disk should hardly show such a large difference in such an essential property. The theory had the advantage, though, of not appealing to a highly unlikely event such as capture, instead envisioning the Moon as a natural consequence of the formation of the solar system, thus perhaps explaining the origin of the other satellites as well.

By the 1950s, each of the three classic theories appeared to have at least one fatal flaw. With no new data to chew on, disagreements over the origin of the Moon had reached a kind of stalemate, just as had the question of the origin of lunar craters before the Ranger missions to the Moon. But, as in that case, new data were about to arrive, eagerly awaited by a new generation of geologists, physicists, and a Nobel Prize–winning chemist.

9

We Go into Space

The scientific study of the Moon began when Galileo turned his telescope upon it. A new era opened in 1961, when President John F. Kennedy announced that an American would go to the Moon. It was clear from Kennedy's speech that one reason for his challenge was to prevent the Soviet Union from gaining an unbeatable lead in space exploration, which the launch of the beep-beeping Sputnik in 1957 had shown was entirely possible. But in a special message to Congress, Kennedy also set out a loftier purpose: "We go into space because whatever mankind must undertake, free men must fully share."[1]

The obvious question was what an astronaut on the Moon should do there. The answer seemed equally obvious, at least to any scientist: collect Moon rocks and bring them back for study. The scientific justification for the expense and risk of the space program would ensure that scientists would play a major role in it. None had more influence than chemist Harold Clayton Urey, a key member of the Manhattan Project. As it had done for the young Shoemaker, reading Baldwin's *The Face of the Moon* had changed Urey's life.[2]

From a small Indiana town, Urey taught school for a year, then entered the University of Montana, where he conducted the first of a lifetime of research projects, this one on the protozoa in a channel of the nearby Missouri River. Urey entered graduate school in chemistry at the University of California at Berkeley, where he published one of his first papers in the *Astrophysical Journal* (on a correction to the ideal gas law), as the chemist would enjoy pointing out to the astronomers who became his colleagues decades later.

After receiving his doctorate in 1923, Urey earned a fellowship to study with the pioneering physicist Niels Bohr in Copenhagen, giving him a foundation in the new quantum theory that most American chemists at the time lacked. By 1931, scientists had recognized that elements can have different isotopes, but not until the next year did James Chadwick discover the neutron, which revealed that these varieties of an element have the same number of protons but different numbers of neutrons. Though scientists had not yet identified other isotopes of hydrogen, Urey and others had proposed that in

Unlocking the Moon's Secrets. James Lawrence Powell, Oxford University Press. © James Lawrence Powell 2023.
DOI: 10.1093/oso/9780197694862.003.0010

addition to common hydrogen of mass 1, a rare stable (nonradioactive) iso-tope of mass 2 existed. Urey devised a clever method that would enrich the amount of hydrogen in a sample enough to bring the second isotope up to the level of detection. He did find it, naming it deuterium and earning him the 1934 Nobel Prize in Chemistry. The story goes that Urey made the dis-covery on Thanksgiving Day but still got home in time to report the feat to his wife and enjoy the family Thanksgiving dinner.

By 1940, scientists had discovered that atoms can split into smaller ones, a process known as nuclear fission, and had established that the isotope of ura-nium that undergoes fission is U-235. They had also established the principle of the chain reaction, in which neutrons released in fission can cause fission in additional atoms, in an accelerating cascade that leads to an atomic explo-sion. U-235 is much less common than U-238, and to build a bomb, U-235 must be separated and enriched. Who better to turn to than Urey? He began work on uranium isotope separation in May 1940, a year and a half before the United States entered World War II. Urey conceived of the idea of using a spinning centrifuge, which separates isotopes according to their mass and remains the favored method for concentrating the lighter U-235. One clue that a modern nation may intend to build a nuclear weapons program is that it purchases high-speed gas centrifuges.

After the war, many American scientists, particularly those who had worked on the atomic bomb, began to fear what nuclear weapons might do to humanity. Only months after the use of the atomic bomb on Japan, Urey wrote in the popular *Colliers* magazine, "All the scientists I know are frightened—frightened for their lives and frightened for your life."[3] Urey began a crusade for world governance of atomic weapons, a cause that led the US House Committee on Un-American Activities to attack Urey and led the FBI to investigate him.

Urey and several colleagues moved from Columbia University to the University of Chicago, where he began to investigate another peaceful application of isotopes: the use of those of oxygen to measure ancient temperatures. O-16 is by far the most common isotope of oxygen, but ones of mass 17 and 18 also occur. Because the lighter O-16 evaporates more easily than the other two, its proportion rises in the atmosphere and in subsequent precipitation, while leaving the source of the evaporation relatively enriched in O-18. The process is thus sensitive to temperature, allowing the ratio of the isotopes to serve as a paleo-thermometer. The oxygen isotope thermometer has revolutionized several fields of geology analogous to the way radiocarbon dating had transformed several other sciences. The oxygen isotope ratios in deep polar ice cores are the principal evidence that twenty-first-century temperatures are the highest in the last several million years and are rising on a human timescale, not a geological one. As we will see, oxygen isotope ratios also bear strongly on theories of the origin of the Moon.

Then came an event that changed Urey's field of interest from the minute scale of isotopes to a colossal question that scientists could not answer by laboratory experiments. In 1949, Baldwin's *Face of the Moon* had just come out. In her interview with Baldwin, Marvin reported on a conversation she had with Cyril Stanley Smith, a professor at the University of Chicago at the time:

> One evening, the Smiths held a cocktail party for faculty members, and Harold and Frieda Urey were among the first to arrive. Harold seated himself comfortably at the end of the couch and Cyril handed him a book saying, "Harold, you had better look at this." Hours later, someone asked Frieda where Harold was: "Isn't he coming?" "Why, yes," she said, "we came together, he has been here all evening." A search located Harold still sitting at the end of the couch totally engrossed in *The Face of the Moon*.[4]

According to his biographers, thus began Urey's "love affair" with the Moon. "Colleagues, or any available listeners, would be treated to monologues, sprinkled with the names of craters and other technical terms, which were impressive though bewildering." This interest broadened and led to Urey's 1952 book, *The Planets*, the beginning of "a sustained, audacious attack on the broader problem [of the origin of the solar system.]"[5]

In science, raising the right questions is as important as answering them, since it is the unanswered ones that fuel further inquiry and lead to improved theories. In this spirit, *The Planets* raised more questions than it answered, but they were the right questions and therefore served to establish a research agenda for the coming space age. Just as Baldwin's book had changed the lives of Urey and Shoemaker, so Urey's book caused many physical scientists to switch fields, which in turn led to advances in many areas of earth and planetary science, advances that would have taken longer without them. Scientists within a field often resent the entry of an "outsider" from another discipline. Urey and his scientific progeny show how shortsighted this is.

The way in which Urey wrestled with the mystery of the Moon's origin from the early 1950s to beyond the Apollo missions shows a great and open-minded scientist at work. He would settle on a preferred theory—fission, co-accretion, or capture—only to change his mind in the face of new evidence or sometimes in response to a persuasive colleague, then change it again. Urey embodied the saying "To be uncertain is to be uncomfortable, but to be certain is to be ridiculous," variously attributed to Socrates, Goethe, and others. Brush thought that Urey "was too ready to change his ideas and was likely to take both sides on an issue."[6] But as the history of science shows, open-mindedness is a virtue; the real risk is premature rejection. That can and has cost decades of progress in science, as witness resistance to the theory of continental drift and others.[7]

In *The Planets*, Urey's hypothesis for the origin and development of the solar system began with a small, dark nebula that collapsed into a star—our Sun—with enough material left over to form a cool, surrounding disk of gas and dust. The disk then fragmented into masses that formed the Sun and planets. This view melded elements from both the nebular and the planetesimal hypotheses. Following Baldwin, Urey accepted that meteorite impact had created most lunar craters.

In 1951, the year before *The Planets* appeared, Urey published an article in which he applied chemical principles to the problem of the origin of the solar system.[8] One of its most salient sentences read, "In fact, the moon and Mars are fossil planets, each giving information about past events." Urey's principal evidence for this view was that, as scientists had long been aware, the pull of the Earth's gravity causes the Moon to bulge slightly toward our planet. That this protuberance still exists after billions of years means that the Moon, unlike the Earth, lacks a hot interior that could have gradually flowed aside to remove the bulge. This line of thinking leads to the "cold Moon" model, which considers it a primordial object, a relic from the earliest history of the solar system, a fossil. In this view, the Moon provides the best way of learning about the early history of the Earth, since plate tectonics, volcanism, and erosion have destroyed the evidence of our planet's beginning.

* * *

The United States responded to the 1957 launch of Sputnik by speeding up the work on rocketry that had begun at the Marshall Space Flight Center in Alabama and at California's Jet Propulsion Laboratory. In 1958, to guide America's voyages on the new ocean of space, Congress established the National Aeronautics and Space Administration, or NASA. President Kennedy's 1961 announcement that the US would land a man on the Moon was not only daring, but, as Kennedy well knew, it would be expensive. Just showing up the Soviets could hardly justify such a vast expenditure of tax-payer dollars. But a scientific goal would put the project in the same realm as other big science projects, such as the development of atom smashers, seen as appropriate government undertakings. With Urey in the lead, scientists were only too happy to flesh out the scientific reasons for sending astronauts to the Moon.

At first, America's voyages into the new ocean of space flopped, some-times literally, as typified by what NASA's historians call the "absolute nadir of morale among all the men at work on Project Mercury," which was the "four-inch flight" on November 21, 1960, of a Redstone rocket booster that barely lifted off the launch pad.[9] The first six flights of the Ranger program also failed, and then came the success of Ranger 7 and its confirmation that lunar craters derived from meteorite impact, as described here in Chapter 7. By now, a crew of handsome, intrepid, all-American and all-male astronauts stood ready, each willing to risk his life to be the first human to set foot on an-other body in the solar system. And some of these brave men would lose their lives in the effort.

Scientific preparations for a Moon landing began at the new Astrogeology Branch of the US Geological Survey in Flagstaff, Arizona, with Shoemaker as its head. He and his team prepared colorful "geological" maps of the Moon based on photographic images, as shown in black and white in Figure 9.1.

In 1966, Lunar Orbiter 2, sent to search for potential landing sites, shot photos and transmitted them back to Earth. Most of the photos were taken looking almost vertically down and gave no sense of the 3D topography of the Moon's surface. One exception was an oblique view of the floor of Copernicus, shown in Figure 9.2. As quoted in *Time* magazine, NASA sci-entist Martin Swetnick called this "one of the great pictures of the century."[10]

A key scientific question was whether the Moon had been born hot or cold. But the answer also bore on whether astronauts could safely land on the Moon. If the Moon had started out hot, volcanoes would have extruded lava onto the surface, which, when cooled and solidified into rock, would bear the weight of a landing vessel and an astronaut. On the other hand, had the Moon formed cold without subsequent volcanism, its surface would have suffered billions of years of continual pounding by meteorites, quite possibly pulverizing everything into a deep and potentially fatal layer of moondust.

Scientists thought that the age of the rocks at the surface might answer the question. Those on a hot Moon would be as young as the youngest volcanic activity, whereas those on a cold Moon could date back to the birth of the

Figure 9.1. Geological map of the Copernicus quadrangle. (US Geological Survey.)

Moon and the solar system itself. But until astronauts landed on the Moon and brought back specimens, surely dating the age of rocks on the surface of the Moon was impossible. Or was it?

Not according to an inventive young scientist named William Hartmann. By the 1970s, Canadian scientists had found several meteorite impact craters on the Canadian Shield, a vast region of Precambrian rock that underlies most of eastern Canada. Hartmann's idea was simplicity itself—if you had the brainpower to conceive it and if you accepted the existence of terrestrial and lunar impact craters. Hartmann counted the number of Canadian craters with diameters greater than 1 kilometer and noted their age range.[11] This allowed him to calculate the rate of impact cratering on Earth: X craters above 1 kilometer in diameter per square kilometer per million years. He then corrected that rate to account for the Moon's lower gravity, which would have attracted fewer impactors, to estimate the cratering rate on the Moon. Using tele-photographs of the Moon, he counted the number of craters larger

Figure 9.2. The floor of Copernicus. (Wikimedia Commons.)

than 1 kilometer in the maria and applied the derived lunar cratering rate to calculate their age, which turned out to be 3.6 billion years, older than any terrestrial rock dated at that time. The result was plausible, as it was less than the measured age of the solar system and roughly in accord with the chronology of the "cold-Mooners." When scientists dated the returned Apollo 11 basalts from Mare Tranquillitatus, they got 3.7 billion years.

Hartmann's age estimate had a wide error range and, even if accurate, did not settle whether moondust might cover the Moon's surface. But when Surveyor 1 lit on the surface of Oceanus Procellarum (the Ocean of Storms) on June 2, 1966, it bobbed up and down in the Moon's lower gravity, then settled to take a "selfie" of its footpad resting in a small dimple (see Figure 9.3). The subsequent Surveyor landers also survived, ending concern about a possible fatal layer of powder.

One of the Surveyors carried an instrument to bombard the lunar soil with subatomic alpha particles. The resulting energy spectrum allowed scientists to measure the amounts of the major chemical elements in the soil, good enough to determine whether it derived from volcanic rock, as the hot-Mooners hoped for and the cold-Mooners bet against.

In 1967, Surveyor 5 set down in Mare Tranquillitatus, soon to host Tranquility Base, the site of Neil Armstrong's dramatic "one small step" two years later, and the alpha-scatterer came on. Analysis showed that the mare surface was basalt, the most common volcanic rock on the Earth. When cold-Mooner Urey learned of the results, he showed himself willing to abandon a favorite theory, saying, "Maybe Mother Nature knows best."[12] Lunar Orbiter, which followed Surveyor, sent spacecraft circling the Moon to photograph its entire surface, searching for the best landing site for Apollo 11. On June 29, 1969, Armstrong and Buzz Aldrin landed to establish Tranquility Base and collect precious Moon rocks, which they brought back to anxiously waiting scientists.

Figure 9.3. Surveyor 1 footpad on the Moon. (NASA.)

Near the fiftieth anniversary of the Moon landing in 2019, Apollo 17 astronaut Harrison Schmitt, the only geologist-astronaut yet to walk on the Moon, praised Armstrong's sample-collecting skills. Schmitt said that even if the Apollo program had ended with number 11, with no additional landings (including his own), the samples that Armstrong collected "would have been enough to forever reshape knowledge of the solar system."[13]

10

Rosetta Stone of the Solar System

To analyze the Moon rocks brought back by the Apollo 11 astronauts, NASA chose a select group of scientists. They and the larger community of lunar specialists no doubt believed that the analysis would answer the hot Moon/cold Moon question and show which of the three classic theories was more apt to be correct. As we have seen, each of the three had advantages and deficiencies. Here is how scientists saw them prior to Apollo.

The fission theory had several defects. (1) In order to fission, the Earth would have needed roughly four times the present angular momentum of the Earth-Moon system. In other words, the proto-Earth would have had to have spun several times faster than today, with no obvious way to slow it down. (2) Even so, a viscous proto-Earth could not have spun fast enough to throw off material the size of the Moon. (3) Once the bulge had broken off and escaped, it would have been inside the Roche limit, where the gravity of the proto-Earth would have broken it into fragments and prevented it from aggregating. (4) Fission would have left the plane of the Moon's orbit close to the plane of the Sun's apparent path through the sky (the ecliptic). Instead, the Moon's orbital plane inclines to the ecliptic by 5.14 degrees (see Figure 1.2). This is a seemingly small amount, but it looms large if you are trying to decipher the Moon's origin. Proponents of fission sought ways around these difficulties, but it was a formidable list.

According to the co-accretion or binary-planet theory, the Moon and the Earth would have had the same original chemical composition and density, which they do not. Co-accretion also failed to explain the angular momentum of the Earth-Moon system.

As for the capture theory, scientists had modified it enough to eventually win the support of Urey and others, but it, too, had serious difficulties. (1) If the Moon came from outside our solar system, it would have had to escape the gravity that held it in its birthplace, then travel through interstellar space on just the right trajectory to be caught by the Earth's gravity. It is much more probable that the traveler would either swerve around the proto-Earth and enter a new orbit around the Sun or else collide with the

Unlocking the Moon's Secrets. James Lawrence Powell, Oxford University Press. © James Lawrence Powell 2023.
DOI: 10.1093/oso/9780197694862.003.0011

Earth. (2) The captured body would have been traveling on a highly eccentric (noncircular) orbit, making the Moon's present nearly circular orbit hard to explain. (3) Capture appears to be a one-off event that does not account for the scores of other moons in the solar system.

* * *

The first returned Moon rocks came from Mare Tranquillitatus, chosen by NASA as the safest place for Apollo 11 to land. By this time, scientists agreed that the mare floors were made of basalt, as Dietz had suggested decades before. One of the first and most important findings from the lunar samples was that the Tranquility Base basalts date to around 3.7 billion years ago, close to Hartmann's earlier estimate. Scientists had established the age of the Earth and the solar system at 4.5 billion years, while the oldest terrestrial rocks dated at the time were around 3.5 billion years old. That the first Moon rocks to be dated were older than the oldest rocks on the Earth suggested that the Moon might indeed be the fossil that Urey had in mind.

The report of the Lunar Sample Preliminary Examination Team, remarkably published only two months after the sample return, described the great age of these first samples as the mission's most "exciting and profound observation."[1] They concluded: "It seems quite likely that if the rocks from Apollo 11 do not take us back to the time of formation of our sister planet [the Moon], then rocks from other regions on the moon will." This was another accurate prediction, as continued exploration and measurement found Moon rocks that date back beyond 4 billion years, with several older than 4.4 billion, getting very close to the age of the solar system itself.

Chemical analysis of the returned samples would be critical, as it should allow scientists to discriminate among the three classic theories. Instead of placing too much stock in the abundance of individual chemical elements, which can vary for any number of reasons, lunar scientists considered them in groups with similar chemical properties. Refractory elements resist melting and evaporation, whereas volatile ones boil and evaporate at low temperatures. The siderophile, or "iron-loving," elements, which follow iron in chemical processes, have a lower abundance in the Earth's crust than in the solar system generally, likely because when the Earth formed, they followed iron into the molten core. Thus, the abundance of these element groups compared with the Earth and the solar system as a whole might reveal whether the Moon has ever melted completely and whether, like the Earth, it is likely to have a core.

The examination team found the returned basalts enriched in refractories such as titanium and zirconium, while volatiles such as lead and bismuth were strongly depleted, as were the siderophile elements platinum, gold, silver, nickel, and cobalt. Another discovery, and one of the most surprising, was the complete lack of hydrated minerals, those that had incorporated water molecules into their crystal structures. Their absence showed, as the report put it, that "[t]here has been no surface water at Tranquility Base since the rocks were exposed" (i.e., for 3.7 billion years). Urey and others had once entertained the idea that the flat maria were sediments deposited on the floors of long-vanished lunar lakes, but the utterly dry samples quashed this no doubt sincerely regretted notion. Perhaps no other feature better revealed the difference between the history of the sterile Moon and our blue and green Earth than their opposite amounts of life-giving and life-sustaining water.[2]

Once scientists got lunar rocks under their microscopes, they found innumerable "zap pits," as they dubbed them, glass-lined cavities ranging in size from a few microns (millionths of a meter) up to millimeters. The impact of tiny particles traveling at cosmic velocities had melted and vaporized the target rock, leaving the microscopic craterlets. As we have seen from Galileo onward, the better the telescopic resolution, the more and the smaller the craters that observers saw. The returned lunar samples showed this to be true even over a size range of 1 trillion times between the micron-sized zap pits and the largest lunar structures, such as the ringed, 1000-kilometer-wide Orientale Basin, part of which peeks out from the Moon's edge (see more on this in Chapter 12).[3]

Scientists interpreted these initial chemical results to mean that the Moon differs enough from the Earth to weaken both the co-accretion and the fission theories, strengthening capture by default. Yet the one-off improbability of capture remained a limitation. Urey got around this difficulty by proposing that the early solar system would have contained many objects of an appropriate size for capture, making the process not so improbable after all. This led to a modification of the original capture theory to assume that a Moon-sized body had gotten inside Earth's Roche limit and broken into pieces, forming a ring from which the Moon assembled.

Proponents of each theory tried to find a way to accommodate the chemical differences but failed to win over those who favored one of the other two. By the early 1970s, the only thing that scientists seemed to agree on was that the Moon was likely as old as the Earth and the solar system. Social scientist Ian Mitroff reported evidence of the lack of consensus in a 1974 book titled

The Subjective Side of Science. He had interviewed forty-two lunar scientists, asking them to evaluate five theories for the origin of the Moon: fission, capture, and three versions of the co-accretion hypothesis.[4] On a scale ranging from 1 ("agree strongly") to 7 ("disagree strongly"), Mitroff found that none got better than a 4. Half the interviewees "expressed little or only moderate interest regarding the origin of the Moon," a sign of how insoluble the old problem had proven. One scientist told Mitroff:

> We've lived with some of these theories for so long that they don't mean much to us anymore. We've heard the same old people spin out the same old cobwebs and speculations for years without adding much to them. For the first time in history we have the hard data, the evidence to test a lot of these things. It's the data that is important now.

* * *

In the years after Apollo 11 (and right up to the present day), scientists continued to study the returned Moon rocks, confirming the early findings of irreconcilable chemical differences between the Earth and the Moon. Yet they got no closer to settling on a single theory of lunar origin. Then, at a 1975 conference, Robert Clayton (1930–2017), who had inherited Urey's lab at the University of Chicago after the great chemist retired, reported a new kind of evidence. His research showed that the Earth and the Moon had identical oxygen isotope ratios.[5] Having focused on the chemical differences between the two bodies, scientists now had to confront a confounding identity on the minuscule scale of isotopes. Remember that oxygen has three naturally occurring, stable isotopes, ones that are neither radioactive nor formed by radioactive decay. Almost 98 percent of oxygen is O-16, with O-17 making up about 0.04 percent and O-18 about 0.20 percent. O-16 is the most abundant of the three because it is a "primary" isotope, made by stars that start out with only hydrogen. The other two are "secondary," in the sense that they require the nuclei of elements above hydrogen, such as helium or nitrogen, to be built first in primary stars and then used to form subsequent elements, in a two-step process.

In trying to understand how the Earth and the Moon could wind up with identical oxygen isotope ratios, scientists needed to know whether those ratios were common to everything in the solar system, in which case the identity would only be expected, or whether they varied from region to region, in which case some process must have equalized the oxygen isotope ratios of the Earth and the Moon. To that end, Clayton and colleagues had also measured the oxygen isotope ratios from different classes of meteorites that were believed to have formed in different regions of the solar system. They found that the ratios in each class differed from each other and from the Earth and the Moon. Thus, a proto-Moon captured from some distant part of the solar system would have been unlikely to have arrived with an oxygen isotope composition identical to that of the proto-Earth, weakening the capture theory. On the other hand, bodies formed in the same region of the solar system from the same material might well be identical down to their isotopes, strengthening the co-accretion hypothesis. In fission, if the proto-Earth had been homogenous, material flung from near its surface to form the Moon might also have left both bodies with the same oxygen isotope ratios.

To sum up where matters stood by the mid-1970s, after a century of scientific effort and an expenditure on the Apollo program of more than $280 billion in today's dollars, scientists were still not sure where the Moon came from. Urey and others had proffered the Moon as the Rosetta Stone[6] of the solar system, yet its chemical and isotopic hieroglyphs had proved as difficult to decipher as the original. But scholars eventually did learn to translate Egyptian hieroglyphs, opening a portal into ancient history. The insight that solved the riddle came from a new way of looking at the problem, when Jean-François Champollion realized that ancient Egyptian had descended from Coptic. Was there a different way of looking at the origin of the Moon that would unveil its history? Before turning to that question, let us return briefly to another: whether the Moon has volcanic rocks.

11

Volcanism on the Moon

From Hooke down to the astronomers of the twentieth century, with few exceptions, scientists regarded lunar craters as volcanic. Then came the close-up images from Ranger 7, which led to a virtually instantaneous paradigm shift as most scientists rejected lunar volcanism and embraced meteorite impact. Yet if the Moon has never had what we would recognize as a volcano, where did the basalt that fills the mare basins come from?

On the Earth, internal heat melts portions of the upper mantle to produce magmas with the chemical composition of basalt. Being less dense than the mantle around them, these magmas rise to erupt as lava on the surface, where they typically build up low, broad "shield volcanoes" like those of the Hawaiian Islands. But the Moon has a second potential source of the energy needed to melt its solid interior, that released by the giant collisions that created the larger lunar craters and basins. If meteorite impact directly produced the basalts of the mare plains, then their age would be the same as or only slightly younger than that of the impact basin that enclosed them. But when scientists dated basalt samples from the giant Oceanus Procellarum, for example, they turned out to be more than a billion years younger than the estimated age of the impact basin itself, one of the oldest on the Moon. Thus, the energy of impact did not produce the basalts of the mare basins; they came much later, from the internal heat of a "hot Moon" after all. Impact might have assisted by fracturing the Moon's surface to provide channels through which these basaltic magmas could later rise to fill the mare basins. This raises the question of whether there might be other features on the Moon that derived from internal melting but could not be detected from telescopes on Earth. The Lunar Orbiter missions, which followed Apollo and photographed 99 percent of the Moon's surface at a resolution of 60 meters or better, turned up several possible volcanic features. Some are quite young, no older than a few tens of millions of years, showing that volcanism has been going on for almost all of the Moon's history.[1] On the Earth, the decay of radioactive elements generates heat that leads to volcanism. But those elements

Unlocking the Moon's Secrets. James Lawrence Powell, Oxford University Press. © James Lawrence Powell 2023.
DOI: 10.1093/oso/9780197694862.003.0012

are rare in the Moon's interior, scientists believe, leaving them to search for an alternative source of the Moon's volcanism.

The lunar "rilles," sinuous, incised valleys that wend their way across a mare plain and vanish, provide one possibly volcanic feature. One of the largest is Hadley Rille, shown in Figure 11.1, named for lawyer and amateur meteorologist George Hadley (1685–1768). He was the first to propose that Earth's rotation plays a role in sustaining the all-important trade winds, a mechanism known as Hadley circulation. Were Figure 11.1 a scene on Earth, it would not occur to any geologist to imagine that Hadley Rille derived from volcanism; rather, it would appear to be a valley cut by a stream that had flowed across the surface, then dried up. But the Moon has never had liquid water on its surface, so volcanism of some sort is the only remaining option to explain the rilles. One hypothesis is that lava flowed down the mare plain and carved the rille—erosion by volcanism. Another is that the rilles are like the lava tubes seen on the slopes of giant shield volcanoes. There lava flows

Figure 11.1. Hadley Rille. The Apollo 15 landing site is just above the center. (NASA.)

down channels and gradually solidifies to build up their sides. A later lava flow can roof over the channel and, when the roof collapses, leave a serpentine channel. In this case, the rille would have been built up, not carved out.

On July 3, 1971, Apollo 15 landed about a kilometer from Hadley Rille. This was the first Apollo flight to bring along a lunar rover, a sort of high-tech golf cart that greatly expanded the area the astronauts could explore. When they reached the edge of the rille and looked to the other side, they saw layers of basalt sticking out through the debris. The 1994 Clementine orbital mission found that the walls of sinuous rilles typically expose such layers of basalt. The balance of scientific opinion now favors the hypothesis that the rilles represent collapsed lava tubes.

The Apollo 15 crew collected a record haul of samples, including one that appeared to be a "basement" rock from the original surface of the Moon. It was an anorthosite, a rock type made mostly of crystals of the mineral plagioclase feldspar. But the "Genesis rock," as scientists prematurely nicknamed it, turned out to be "only" 4.1 billion (plus or minus 0.1 billion) years old. Another group of anorthosite samples gave 4.46 billion (plus or minus 0.04 billion) years, very close to the age of the solar system itself. Thus, some of the anorthosite of the lunar highlands does appear to date back nearly to the genesis of geologic time.

Large shield volcanoes consist of low-viscosity lavas that, rather than piling up to form cones, flow outward for long distances. Thus, these volcanoes have gentle slopes. The steep-sided stratovolcanoes, in contrast, are made of alternating layers of hard lava and fragments of magma that were blasted into the air, where they froze and fell back to Earth nearby and were then covered by the next lava flow, as we see in iconic volcanoes such as Fuji, Rainier, and Vesuvius.

Neither type of volcano exists on the Moon. Lunar lavas are extremely low in volatile elements and have lower viscosity than any terrestrial magma, allowing them to flow for many kilometers and leave a nearly flat mare plain. They have virtually no water to create steam and propel volcanic eruptions, and in any case, the Moon's internal pressures are too low to blast out material that could settle back to form a volcanic cone. Any material that did happen to be hurled above the surface would have emerged under low gravity, in a vacuum, and would have scattered far and wide from its original vent.

After that, it comes as a surprise to discover that the Moon does have rare domes and hills that appear to derive from volcanism. But instead of resembling a lofty Kilauea, they are only a few hundred meters high and a few kilometers long. The Marius Hills, located in Oceanus Procellarum and shown in Figure 11.2, number around three hundred and cover an area roughly equivalent to that of a medium-sized US state. Their domelike shapes show that they consist of an anomalous type of lava too viscous to flow easily.

On Earth, volcanic rocks show the same wide range of composition as more coarse-grained ones. A granular type called gabbro has roughly the same mineralogy and chemistry as basalt, while coarse-grained granite corresponds to a light-colored lava called rhyolite. The latter are high in silica, which increases their viscosity, or stiffness. Some twenty of the returned Apollo samples turned out to have the composition of granite, showing that such silica-rich rock types are present on the Moon. On the Earth, these stiff magmas can plug a volcanic vent and build up pressure that leads to an explosion, like the one that destroyed Mount St. Helens in 1980. Alternatively, if such turgid magmas reach the surface, they can form domes like those

Figure 11.2. Overhead view of the Marius Hills from the Lunar Reconnaissance Orbiter. The 41-kilometer crater Marius is at the right. The hills rise above the surface; the crater is indented into it. (NASA.)

of the Marius Hills or the 3-billion-year-old Gruithuisen Domes, shown in Figure 11.3. NASA has funded a mission that will land a spacecraft and rover among the domes and analyze the rocks with a nuclear spectrometer, a much-advanced version of the alpha-scatterer whose results surprised Harold Urey. Two professors at the University of Central Florida, planetary scientists Kerri Donaldson Hanna and Adrienne Dove, will direct the mission.

Thus, although the Moon has no volcanic craters, it does have igneous (once molten) rocks. These include the mare basalts, the anorthosites that make up the lunar highlands, and the rare rhyolite hills. These must derive from the Moon's internal heat, left over from its beginning.

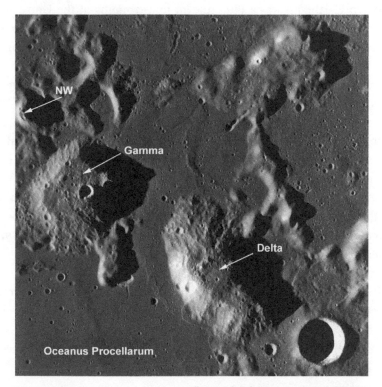

Figure 11.3. The Gruithuisen Domes. The width of the area shown is about 64 kilometers. Gamma Hill is about 20 kilometers long. The crater at the lower right is Gruithuisen B. Scientists believe they are made of viscous rhyolite. (NASA.)

12

Giant Impact

In 1974, the International Astronomical Union met at Cornell University to discuss the moons of the solar system. The chairman of the organizing committee observed that the conference was overdue, because "Until very recently, satellites were the neglected children of the solar system family."[1] Only two papers dealt with the origin of the Moon. William Hartmann, who had used crater counts to accurately predict the age of the lunar maria, coauthored one of them.

When Hartmann was a PhD student in astronomy at the University of Arizona in the early 1960s, he worked on a "rectified lunar atlas." His technique was to project photos of the Moon onto a globe, then to photograph them as though they were seen from above. This allowed him a better view of features on the outer "limbs" of the Moon, where the nearside and the farside meet. One feature now visible was the giant "ring formation" Mare Orientale, shown in a 1967 Lunar Orbiter photograph in Figure 12.1. Of course, Hartmann's image was neither as sharp nor as inclusive as that one, but it was enough to allow him to recognize Orientale as a huge impact basin.

One can only imagine that had the early astronomers been able to see Orientale, they would have given up the idea of lunar volcanism at least for the large, ringed basins. For us, it is impossible to look at the bullseye of Orientale without seeing it as the result of a colossal impact. As Hartmann remembered decades later:

> The perfectly preserved, 1000-km-scale impact structure, hitherto unrecognized on the east limb of the Moon, set me wondering what were the largest impacts that had ever occurred on Earth or in the Earth-Moon system, what were their effects, and what other gross phenomena had been missed in the 1960s' quest for ever greater lunar detail?[2]

We might rephrase: If an object large enough to have created Mare Orientale had been present in the early solar system, other similar-sized bodies must also have been around. Where did they go?

Unlocking the Moon's Secrets. James Lawrence Powell, Oxford University Press. © James Lawrence Powell 2023.
DOI: 10.1093/oso/9780197694862.003.0013

Figure 12.1. The 1,000-kilometer-wide ring formation Mare Orientale from Lunar Orbiter 4. (Wikimedia Commons.)

Hartmann and Don Davis, an astronomer and member of the team that brought the nearly fatal Apollo 13 mission home safely, titled their 1974 conference paper "Satellite-Sized Planetesimals and Lunar Origin."[3] They followed up Urey's notion that many moon-sized bodies might have been present in the early solar system, making a capture event less ad hoc and perhaps even probable. Hartmann had noted in a 1972 paper that studies of the rate of lunar cratering and the presence of huge impact structures like the Orientale Basin were evidence that planetesimals larger than anyone had previously conceived had menaced the early solar system.[4] "During its accretion," he wrote, "the Earth must be viewed as surrounded by numerous smaller bodies all in solar orbit."

Using mathematical equations, Hartmann and Davis set out to calculate how those smaller bodies would have collided with each other and stuck together to build larger ones, which collided with still larger ones, and so on.

They found that the result would be a proto-planet with the approximately 6,000-kilometer radius of the Earth and a nearby second body with a radius of from 500 to 3,000 kilometers, the latter about the size of Mars. After these would come dozens of objects with radii between 100 and 500 kilometers. Hartmann and Davis found that if an object with a radius of 1200 kilometers, traveling at a cosmic velocity of 13 kilometers per second (30,000 mph), had struck the proto-Earth, the collision would have had more than enough energy to throw a mass the size of the Moon into orbit. The model had a simple and inexorable logic: however large the primary planet, a second-largest body must have been present in the same region of the early solar system, raising the likelihood, if not the inevitability, of a collision between behemoths.

The concept of random but predictable events led Hartmann and Davis to an "important philosophically satisfying aspect" of their hypothesis: it could explain the baffling diversity of the many moons of the solar system. They resemble an assortment of apples and oranges—and pears, lemons, bananas, cherries, and pick your fruit—making it hard for a single theory to account for them all. Jupiter and Saturn have many tiny moons that circle their host planets like miniature solar systems, some in "retrograde" orbit, in the opposite direction to the spin of their respective planets. Uranus has a hodgepodge of moons at different distances and with different properties. Earth has its one relatively huge moon, while Mars has only tiny Deimos and speedy Phobos. Venus and Mercury have no moons.

Thanks to the use of computers to analyze several years' worth of telescopic images, scientists continue to discover new moons around the planets. In 2018, Scott Sheppard of the Carnegie Institution announced the discovery of twelve new moons of Jupiter, bringing the total to seventy-nine.[5] He went on to find twenty new moons around Saturn, for a total of eighty-two, breaking Jupiter's just-set record. Seventeen of the newly discovered moons of Saturn have retrograde orbits, while the other three are prograde. Upcoming missions to the two planets will surely increase both numbers. It is hard to imagine anything further from the predictions of the nebular hypothesis than this bizarre assortment of oppositely orbiting moons.

Hartmann and Davis went on to ask a set of questions that illustrate the versatility of their theory in determining the fate of satellites, a case study in how questions drive science:

Does the second-largest planetesimal in each system hit the planet after 10^7 years or 10^8 years? Is it large or small? Does it hit the planet dead center?

Retrograde? A glancing blow prograde? Or is it captured? Or is it destroyed by a planetesimal-planetesimal collision so that it has no appreciable effect on the planet other than to produce many small craters? Or does it hit a pre-existing satellite of the planet, perhaps converting it to several small satellites? Only one of these kinds of fates can befall the second-largest planetesimal. And this fate, the product of chance encounters, may determine whether the planet acquires a tilted axis, a massive circumplanetary swarm of dust, a captured satellite, or perhaps loses a larger satellite, gaining small fragmentary satellites.

* * *

The giant impact hypothesis, as it came to be called, can do more than account for the peculiarities of the moons of the solar system; it may explain the even greater differences among the planets themselves. One of the most peculiar is that each planet's axis tilts at a different angle—again opposite to what the nebular hypothesis would predict. Mercury has almost no axial tilt, while at nearly 90 degrees, Uranus lies on its side. The Earth's axis leans at 23.5 degrees on average. Mars has an axial tilt of 25 degrees, close to the Earth's, producing seasons on the Red Planet like those on ours.

The different axial tilts are just one of many strange features of the terrestrial planets. I once compared them to the barflies in the *Star Wars* cantina, each weirdly different.[6] Start with Mercury, nearest the Sun, at the farthest stool. It is the smallest planet, only about one-twentieth the mass of the Earth, and has no moon, no atmosphere, and the highest density of any object in the solar system. Mercury rotates so slowly on its axis that a day is twice as long as a year. On the next stool sits Venus (aka the Evening Star), comparable in size to Earth but with a dense, suffocating, blistering atmosphere and, like Mercury, lacking a moon. All the other planets orbit the Sun counterclockwise as viewed from the Sun's north pole—prograde—but Venus orbits clockwise, or retrograde. The Earth has a life-giving atmosphere and a large and iron-poor moon. Mars is a little more than half the size of the Earth, has tiny Phobos and Deimos, and has barely any atmosphere. Out past Mars, on the way to Jupiter's stool, where we expect to find a planet, we come across instead a band of asteroids. The gravity of massive Jupiter likely prevented them from accreting into a proto-planet.

Thus, the more we know about the solar system, the less it resembles the orderly outcome of the nebular hypothesis. Instead, it is an almost random collection of objects of different sizes and compositions doing different things, the product of a chaotic beginning. The only characteristic the planets and moons seem to have in common is that all obey Newton's law of gravitation.

As he introduced his giant impact hypothesis at the time of the 1974 conference, Hartmann, still only in his twenties, found himself facing an audience of much more senior and well-published lunar scientists, most of whose ideas he was about to reject.[7] One was Harvard astrophysicist A. G. W. Cameron (1925–2005), at the time perhaps the foremost authority on the origin of the solar system.[8] He was "a large, imposing man, who talks in a slow, pontifical fashion that might indeed strike fear into the heart of a young astronomer." As Hartmann later remembered, when Cameron's hand went up during the question period, "I feared the worst. But Cameron said, 'We're working on the same idea, and we're coming to the same conclusion.'" Cameron and colleague William Ward had also arrived at the hypothesis of a giant impact and found that to spin off a body the size of the Moon, the impactor needed to have been as large as Mars. Such a collision would have generated temperatures as high as 8,000 degrees C, enough to vaporize the outer part of both the impactor and the proto-Earth. The exploding gases would have lifted enough material into orbit to make the Moon and provide the right energy "kick to the system," as Cameron put it, referring to the angular momentum of the Earth-Moon system.

The giant impact theory soon supplanted the three classic theories, while incorporating elements of each. By the early 1970s, scientists had modified the capture theory to include the breakup of the captured body into planetesimals, an early stage of giant impact. As in the co-accretion model, the proto-Earth and Mars-sized impactor were believed to have roamed the same region of space, both having come from the same part of the primordial dust cloud. And at least part of the material that made the Moon came from the Earth, though what proportion would become a major test of the giant impact theory. The fate of the three classic theories shows how in science, the question is not so much whether a given theory turns out to be wrong but whether its testing helps to advance science and strengthen its replacement. For the nebular hypothesis and each of the three classic theories of lunar origin, the answer is a resounding yes. But still, all eventually failed. Would giant impact fare any better?

13

Green Light for Giant Impact

By this point in our story, we have learned that, by and large, "There is nothing new under the Sun," from Ecclesiastes 1:9 and as Richard Proctor put it when he discovered that the "meteoric theory" was not original with him. Thus, it comes as no surprise that scientists beginning with Chladni in 1798 had conjectured that impacting and aggregating meteorites had built the moons and planets of the solar system, the core idea of giant impact. Gruithuisen, Proctor, Gilbert, and Wegener all had the same notion, but none was able to develop it further. The first to do so was Reginald Daly (1871–1957), like Cameron a Canadian-born scientist who became a professor at Harvard University. In a 1946 article titled "Origin of the Moon and Its Topography," Daly made short work of the debate over the origin of lunar craters. He paid homage to Gilbert and noted the inconsistency between the observed features of the Moon and the products of volcanism.[1] He noted that scientists had found "30,000 visible pits" on the Moon's surface and had concluded that all are volcanic vents. Instead, he said, "the large, dominant depressions on the moon [are] primarily the results of as many [meteorite] infalls."

Daly examined variations on the classic theories, finally settling on one that involved capture. In his preferred model, the Moon was a "planetoid which, after striking the earth with a glancing, damaging blow, was captured." That was close indeed to what Hartmann and Davis, who did not know of Daly's paper, proposed at the 1974 Cornell conference. The collision would have been particularly energetic, Daly wrote, "[i]f the 'planetoid' had a large fraction of the moon's present mass" and if at the time the Earth's surface had been molten. The explosion would "drive out terrestrial material into a limited belt ringed about the planetoid," from where it would be "gravitatively aggregated by the pull of a master fragment or captured 'planetoid' to make the substance of our moon, and the somewhat diminished earth felt a prolonged rain of other earth-fragments, large and small." Daly listed observations that the hypothesis of an initial collision and the subsequent rain of "earth-fragments" would explain, including the Moon's mass, density, density distribution, overall shape, albedo (reflectivity), topography, central

Unlocking the Moon's Secrets. James Lawrence Powell, Oxford University Press. © James Lawrence Powell 2023. DOI: 10.1093/oso/9780197694862.003.0014

peaks, and "rays" that radiate from the larger craters. However, in such a violent process, Daly wrote, "[i]t would be hardly possible to tell the proportions of the moon's mass assignable to fragmented planetoid and fragmented earth respectively." This has become the key question regarding the giant impact hypothesis: how much of the material in the Moon came from the Earth and how much from the impactor? During the twenty years after Daly's article, it was cited only once in a peer-reviewed journal—and dismissed.[2]

For a while, the Hartmann-Davis giant impact model also failed to gain traction. Hartmann had been confident that the merit of his argument would carry the day, but "[n]o one had paid much attention to our 1975 paper [from the 1974 Cornell conference] or Cameron's in 1976," he remembered. "I had been naive: I had thought that all you had to do was write a paper, and that was that—it would sink or swim on its own merits. But that's not so. You have to push a new idea."[3]

During the decade after the Cornell conference, Hartmann worked mainly on the 1971 Mariner 9 mission to Mars, which became the first spacecraft to orbit another planet. It showed the Red Planet covered with craters, as well as putative river valleys and a giant extinct volcano. In 1974, NASA's Mariner 10 flew by Mercury and sent back sharp black-and-white images. Until then, planetologists had been unable to discern the surface features of Mercury because of its small size, distance from Earth, and nearness to the blinding Sun. At first, the incoming Mariner images were no sharper than those seen through telescopes, but as the spacecraft bore down, astonishing features began to appear. The most startling may have been an impact basin 1,300 kilometers in diameter, larger than either Orientale or Mare Imbrium. As shown in Figure 13.1, Mercury appears to be at least as densely cratered as the Moon. These findings surely tilted the minds of scientists toward the giant impact hypothesis.

* * *

The next conference on the origin of the Moon took place at Kona, Hawaii, in 1984. Even by this time, ten years after the end of the Apollo program, no consensus on the Moon's origin had emerged. Urey's old joke that science had proven that the Moon does not exist still had an uncomfortable and embarrassing ring.[4]

In preparation for the Kona conference, Hartmann tried to arouse interest in giant impact among his fellow scientists in Tucson, where he worked at the Planetary Science Institute. But as he remembered, geologists

Figure 13.1. Mercury from Mariner 10, 1974. (Wikimedia Commons.)

objected because giant impact was catastrophic and thus violated uniformitarian doctrine.[5] Colleagues counseled Hartmann that "[w]e needed to exhaust all non-catastrophist processes first, before considering giant impacts." Here we see how earth scientists extended the false, or at best only partially true, dogma of uniformitarianism outward from the Earth to the Moon and the solar system. But, Hartmann recalled, "When I went to the planning session to look over the abstracts for the proposed papers, I found, to my amazement and joy, that eight or ten of the abstracts—independently of each other—were about the impact idea." His notion had caught on after all.

Hartmann's Kona conference paper included a section titled "Stochastic Does Not Equal Ad Hoc," stochastic being science-speak for random.[6] The point, he said, was that "[a]lthough we cannot predict when or where a catastrophic event will happen, we can be sure that catastrophic impact events were happening all over the solar system, and therefore we shouldn't rule one out on the basis that it is ad hoc." This was the same point that Urey had made years earlier to support the capture theory.

Hartmann distinguished between events that "are class-predictable but not event-predictable: i.e., we believe the class of events occurred, but we cannot determine times and magnitudes of individual events." His example of a "class-predictable" event was the "probable Cretaceous-ending asteroid impact." "Scientists," he said, "should have pursued the geologic and climatic consequences of these class-predictable events instead of waiting for iridium-rich layers [impact indicators] to take us by surprise." Thus, he attributed to impact two of the most momentous events in the history of the solar system: the giant impact that created the Moon and the collision that caused the death of the dinosaurs.

The Kona conference gave a rare opportunity for proponents of the three classic theories to hear directly from each other, with "experts on all sides so that no one got away with anything," as one researcher told *Science* reporter Richard Kerr.[7] Kerr's attendance reveals an advantage of scientific conferences: they attract skilled reporters who can pull everything together and explain the main findings and their significance to a general audience, as Kerr also did for the Alvarez theory of meteorite impact as the cause of dinosaur extinction and, at least in its early days, for the Younger Dryas impact hypothesis.[8]

By the time of the conference, scientists had come to believe that an ocean—of magma, not water—had once covered the surface of the Moon down to a depth of 200 kilometers. They came to this discovery through a model of scientific reasoning. Kepler had called the dark areas of the Moon the maria, the name by which we still know them. But much of the nearside includes bright, light-colored regions, which he called the terrae. These lunar highlands are more heavily cratered than the maria and therefore have been around longer. When samples of highland rocks reached Earth via Apollo, scientists found them enriched in a common terrestrial mineral called plagioclase feldspar, which, as discussed earlier, makes up the rock anorthosite. Plagioclase is a relatively light mineral—both in color and in density—that would tend to float to the top of a body of magma and concentrate there. This gave rise to the view that the Moon, as Daly had envisioned, had once melted down to a depth of several hundred kilometers, allowing the less dense plagioclase to rise to the surface and coagulate as a mineralogical foam that solidified to create the highlands.[9] Scientists had puzzled over the source of the energy needed to melt so much of the Moon, but a highly energetic giant impact and the continuing bombardment could have supplied it. The magma ocean concept thus came to be seen as supporting giant impact.

The most important chapter of the book that emerged from the Kona conference was its concluding one, "Moon over Mauna Loa: A Review of Hypotheses of Formation of Earth's Moon," by John Wood of the Harvard-Smithsonian Center for Astrophysics.[10] Wood gave letter grades to the three classic theories plus giant impact and a new one, capture and disintegration. All except giant impact got at least one F. It did get three "incompletes."

As Kerr summed up the Kona conference, the giant impact hypothesis had "breathed new life into a long-stagnant field."[11] But would the resuscitation last, or would it represent only the temporary life support that had sustained each of the three classic theories but eventually run out?

14

Mother of Selene

By the end of the 1984 Kona conference, the opinion of lunar scientists had begun to swing in favor of giant impact. And this was a good thing, for without a new theory, they would have had to accept that the billions of dollars spent to send astronauts to the Moon and back and the wealth of new information from the study of the returned Apollo samples all had brought them no closer to unlocking the Moon's secrets. By the time of the next major conference, held in Monterey, California, in 1998, giant impact had become broadly accepted, and attention had turned to how the colossal collision might have led to the observed characteristics of the Moon.

In the years between the two conferences and on to the present day, scientists explored giant impact along two main tracks: (1) testing it against the measured geochemistry of the Earth and the Moon, in particular their comparative isotope ratios, and (2) performing computer simulations of a collision between two large objects in space. In effect, the geochemistry studies provided the facts that needed to be explained, and the simulations tried to come up with a set of starting conditions and a process that could do so. Each track grew steadily more sophisticated as analytical precision and computer power improved.

By the beginning of the twenty-first century, the simulations were showing that most of the material that wound up in the Moon had come from the impactor. The principal thrust of the geochemistry track was to explain how then to account for the identical oxygen isotope ratios of the Earth and the Moon, which Clayton and colleagues had reported in 1975. There were two possibilities. (1) Before they collided, the proto-Earth and the impactor already had the same isotope ratios, which the Earth and the Moon retained post-impact. But if the impactor had come from some other region of space, which seemed likely, for the two bodies to have had the same pre-collision isotope ratios would have been an extraordinary coincidence, never a scientist's first choice. (2) The proto-Earth and the Moon had started out with different oxygen isotope ratios, but they and everything else became homogenized in the giant cloud of white-hot vapor that resulted from the impact. This

Unlocking the Moon's Secrets. James Lawrence Powell, Oxford University Press. © James Lawrence Powell 2023.
DOI: 10.1093/oso/9780197694862.003.0015

possibility also had problems, as isotopes have such similar properties that even when subjected to extreme temperatures, their proportions might have been left virtually unaffected.

Remember that Clayton and colleagues had reported that the oxygen isotope ratios of the Moon and the Earth were identical within the limits of analytical precision at the time. Now jump ahead twenty-five years, one human generation but a tenfold improvement in measurement precision. By this time, it had become apparent that the androgynous "impactor" needed a memorable name. British scientist Alex Halliday proposed Theia, the mother of the Greek goddess of the Moon, Selene, and it caught on.[1]

In 2001, U. Wiechert and colleagues analyzed the oxygen isotope ratios of thirty-one samples returned by the Apollo missions.[2] They found that these lunar rocks had the same oxygen isotope ratios as the Earth to within 16 parts per million. This virtual identity could have resulted, they thought, if Theia had been roughly the size of Mars *and* if both the Earth and Theia had started out with nearly identical oxygen isotope ratios. A 2016 analysis lowered the near identity to 5 parts per million, requiring an even more unlikely coincidence.[3]

The oxygen isotope ratios of so-called lunar meteorites also turned out to be identical to those of the Earth.[4] These are meteorites that were blasted off the Moon by impact and launched into space, from where they eventually traveled to Earth. In 1982, scientists discovered the first of these cosmic voyagers lying on the Antarctic ice, where dark meteorites stand out starkly. No one at the time imagined they had come from the Moon. But when the Antarctic meteorite collection arrived at the Smithsonian Institution for study, geochemist Brian Mason recognized that one of the specimens resembled the Apollo samples more than it did other meteorites. Soon its chemistry, mineralogy, and isotope ratios confirmed that this meteorite had indeed come from the Moon. This suggested that other Moon rocks likely sat unrecognized in museum and university collections, and by 2020, scientists had identified nearly five hundred of them.[5] These specimens have the advantage of coming from random locations on the Moon, including the farside, whereas the Apollo samples were all collected in a relatively small area on the central nearside at sites selected more for the safe return of astronauts than for science. Thus, the lunar meteorites are more representative of the Moon overall, giving even more weight to the identical oxygen isotope ratios of the Earth and the Moon. Scientists have also identified meteorites derived from impact on Mars: Martian meteorites. But their oxygen isotope ratios differ significantly from those of the Earth and the Moon. So do those of meteorites believed to have come from the large asteroid Vesta. Here is a conundrum: the oxygen isotope ratios of the Moon and the Earth are indistinguishable down to the level of a few parts per million, yet they differ from those of Mars and Vesta, which in turn differ from each other. This makes it seem that different regions of the solar system likely have different isotope ratios, underscoring the close genetic connection between the Earth and the Moon.

Scientists went on to discover that not only oxygen but several other elements have identical isotope ratios in both the Earth and the Moon. These include chromium, potassium, silicon, titanium, tungsten, and vanadium. These varied elements, together with oxygen, differ radically in their chemical properties. Oxygen is practically the definition of volatility, while titanium and tungsten are the prototypes of refractory elements that melt only at high temperatures. As an article in *Nature* put it, "For the impactor and the proto-Earth to have the same oxygen isotope ratio is unlikely, but for them also to have the same tungsten isotopic composition is highly implausible."[6] For tungsten, substitute the element of your choice from the list above.

By 2020, forty-five years had passed since the discovery that the Moon and the Earth had identical oxygen isotope ratios. The precision of isotope measurement had increased greatly, but the baffling identical ratios had persisted. Then Erick Cano of the University of New Mexico and fellow scientists came up with a new way of looking at the problem. They reported new analyses of the oxygen isotope ratios of several dozen lunar and terrestrial samples and found, as had previous studies, that the *average* ratio of samples from the Moon was identical to the *average* of samples from the Earth.[7] But they found something new: *individual* samples of different lunar rock types had distinctly different oxygen isotope ratios from each other and from the Earth, well beyond the range of the imprecision in the method. In other words, the ratios of individual lunar samples were *not* identical to those of the Earth overall. It was only when scientists averaged the results and thus removed the individual differences in the lunar samples that the oxygen isotope ratios of the Earth and the Moon matched. Cano and colleagues went on to show that the different lunar oxygen isotope ratios are the result of mixing of two components, the essence of the giant impact hypothesis. Samples brought to the lunar surface from deep in the Moon's mantle have isotope ratios that most closely match those deduced for Theia. Lunar rock types that were believed to have condensed from the primordial impact vapor, and which now make up the Moon's crust, have the greatest fraction of material from the proto-Earth and the lightest oxygen isotope ratios.

These new results showed that before the collision, Theia and the proto-Earth had different oxygen isotope ratios that were not completely homogenized in the impact vapor cloud. This suggested that Theia may have come from a different region of the solar system from the proto-Earth, thus avoiding the unsatisfying appeal to coincidence.

But the Moon always seems to have one more surprise. In this case, as identical oxygen isotope ratios in the two bodies came to be seen as less of a problem, the identity of the tungsten isotope ratios came to be seen as a greater one. These isotopes are more complicated to interpret because in the early solar system, the proportion of one of the tungsten isotopes increased due to radioactive decay of an isotope of the element hafnium. Thus, the tungsten isotope ratios in the early Moon and Earth depend not only on the starting material but also on this secondary process of radioactive decay. Although difficult to interpret, the complications turn out to be a blessing, as they allow the tungsten isotopes to provide a more rigorous test of whether Theia could have had an Earth-like isotopic composition.

Applying this test, a group led by Rebecca Fischer at Harvard found that the probability that the Moon inherited its Earth-like tungsten isotope ratios from Theia is minuscule.[8] If most of the Moon's tungsten did not come from Theia, then it is likely that the other stable (nonradioactive) isotopes did not, either. This implies that Theia and the proto-Earth started out with different chemistries but had them homogenized in the fiery vapor cloud that followed impact, leading to virtually identical average oxygen isotope ratios in the Earth and the Moon.

15

Simulating Giant Impact

Following the 1984 conference, Cameron and colleagues continued to sim-
ulate giant impact, using a method known as smoothed-particle hydrodynamics, or SPH. This technique models the proto-Earth and the impactor
each as a cluster of individual particles, then uses sophisticated mathematical equations to follow the particles forward as they collide.[1] Among the
most important of these is the equation of state, which expresses how the
pressure, temperature, and volume of a substance are related. Thanks to the
ever-expanding power of computing, SPH has become a standard, even rou-
tine, tool of analysis in astrophysics. From the few thousand particles that
computers could handle in the 1970s, the number has now grown to more
than 1 million. Scientists today use the method to model not only the impact
of objects in space but also such gargantuan processes as the formation of
stars and galaxies, the collisions of stars, and the explosions of supernovae.

Cameron and coauthors wrote a series of five articles under the general
title "The Origin of the Moon and the Single-Impact Hypothesis." The first
appeared in 1986 and the last in 1997. Because of insufficient computer
power, the initial version had to make the simplifying assumption that both
colliding bodies had the composition of granite. But even this artificial
starting point showed that under certain conditions, "[t]he formation of a
pre-lunar accretion disk [by giant impact] is almost straightforward."[2] This
would be the case if the impactor had:

- About the same velocity in space as the proto-Earth, 10 or more
  kilometers per second.
- A trajectory that caused it to barely graze the proto-Earth.
- A size roughly that of Mars at about 3,400 kilometers in radius, roughly
  half the radius and one-tenth the mass of the Earth.

Nothing about these requirements is special. They would have been met
many times over in the early solar system. Figure 15.1 shows the result of one
of these early simulations, assuming the three conditions above.[3]

Unlocking the Moon's Secrets. James Lawrence Powell, Oxford University Press. © James Lawrence Powell 2023.
DOI: 10.1093/oso/9780197694862.003.0016

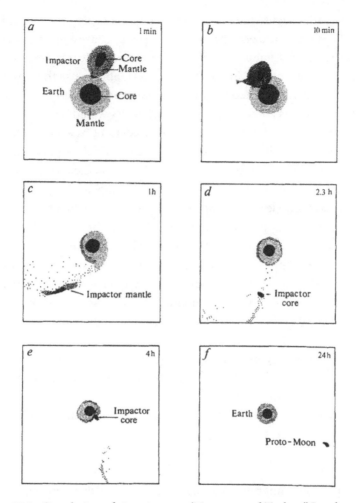

Figure 15.1. Simulation of giant impact. (Newsom and Taylor, "Geochemical Implications.")

By the time of impact, shown in Figure 15.1*a*, both bodies had already developed a metallic core and a silicate mantle. The time elapsed is shown in the upper right in each box. After the collision (*a*), the impactor has begun to spread out in space at (*b*). By the time of (*c*), gravity has caused the debris from the impact to form clumps that would become the Moon. At (*d*), the impactor's iron core has separated from its silicate mantle, and then, at (*e*), the core accretes to the Earth. By twenty-four hours after impact (*f*), a lump of silicate with roughly the mass of the Moon is orbiting the Earth. In this model, 70 percent to 90 percent of the Moon would have come from Theia.

In 2001, Robin Canup of Boulder's Southwest Research Institute (SRI) and Erik Asphaug of the University of Arizona, able to use thirty thousand particles for SPH, presented what came to be called the canonical model of giant impact, the version against which scientists would test new evidence and models.[4] It was similar to Cameron's model but based on much more data: a Mars-sized Theia traveling at a low cosmic velocity had struck the proto-Earth at an angle of about 45 degrees. It, too, predicted that most of the material in the Moon would have come from the impactor.

Canup has been a leader among the new generation of lunar scientists who apply the ever-rising power of supercomputers to the origin of the moons and planets. She received her PhD in astrophysics from the University of Colorado, joined SRI in 1998, and in 2003 won the Urey Prize of the American Astronomical Society for "outstanding achievement in planetary research by a young scientist." Showing that childhood ballet lessons had paid off, one week after finishing her dissertation, Canup danced the lead role in *Coppelia* with the Boulder Ballet. She stopped dancing at age thirty-five, noting, "By that age you're an old dancer but a young scientist."[5] As Canup summed up in a 2004 article, "Overall, perhaps the greatest shift in thinking that has arisen from the past decade of lunar origin studies has been the realization that the impact production of satellites appears an efficient and probable event during planetary accretion."[6]

* * *

As the simulations improved, they continued to show that the original Earth-Moon pair would have had several times the combined angular momentum that the two have today. This presented a formidable difficulty for any theory of the Moon's origin, as angular momentum is conserved and cannot be lost. It can, however, be transferred from one body to another. This happens moment by moment as the tidal pull of the Moon tugs at the Earth and brakes its rotation rate by 66 nanoseconds per day. To conserve the combined angular momentum of the pair, the Moon recedes from the Earth at a rate of about 3.8 centimeters per year.

In 2012, Matija Ćuk and Sarah Stewart illustrated a transfer process that could have lowered the angular momentum of the primordial Earth-Moon pair to that observed. Both have earned the Urey Prize from the American Astronomical Society. Ćuk received his PhD from Cornell and is a principal investigator at the Carl Sagan Center of the SETI (Search for Extraterrestrial Intelligence) Institute. Stewart received her undergraduate degree from Harvard and her PhD from Caltech. She is a professor at the University of California at Davis. In 2018, she received a MacArthur Fellowship for "advancing new theories of how celestial collisions give birth to planets and their natural satellites, such as the Earth and Moon."

The Ćuk and Stewart model began with a proto-Earth spinning two or three times as rapidly as today, the result of many glancing blows in the cosmic pinball machine prior to giant impact.[7] Its more rapid spin would have allowed a much smaller Theia—having as little as 2.5 percent of Earth's present mass—to produce a Moon only 8 percent of which had come from Theia and the rest from Earth's mantle, whose composition would then dominate in the Moon. This would go a long way toward explaining the isotopic identities, but the Earth-Moon system would still wind up with too much angular momentum. To show that this might not be an insurmountable obstacle, Ćuk and Stewart invoked a process that goes by the formidable name "evection resonance." It can reduce the angular momentum of the Earth-Moon system by transferring it to the Earth's orbit, while keeping the total the same. Explaining this complicated process is beyond the scope of this book; suffice it to say that it makes angular momentum less of a barrier to giant impact and allows a greater range of possibilities for the size and composition of Theia.[8]

In a companion paper, Canup accepted the idea that evection resonance with the Sun could have reduced the angular momentum of the early Earth-Moon system. But instead of the canonical Mars-sized impactor or the tiny one suggested by Ćuk and Stewart, Canup, modeling with three hundred thousand SPH particles, proposed that both the target object and Theia had several times the mass of Mars.[9] Her model had the advantage that the closer Theia and the proto-Earth were in size, the more their mixing in the aftermath of giant impact would have left them with the same isotope ratios. In these two papers, we see lunar scientists adjusting the canonical model to better explain the isotope ratios and angular momentum of the Earth-Moon pair. The papers did not convince impact specialist H. J. Melosh (1947–2020) that either was the solution to what he labeled "the isotopic crisis."[10] Melosh received his PhD in physics and geology from Caltech and like Shoemaker, earned both the G. K. Gilbert Award of the Geological Society of America and the Barringer Medal of the Meteoritical Society. He noted that although some of the simulations in the two studies did reproduce the isotopic identities and angular momentum, most did not. As he put it, "A bit more mass to the projectile, a slightly different impact angle or velocity, and the isotopic similarity disappears." He concluded, "In spite of all the progress we have made since the first Apollo landing on 20 July 1969, the very existence of our Moon still conceals mysteries that await solution."

* * *

One example of different starting conditions in a simulation was to assume that at the time of impact, an ocean of magma hundreds of kilometers deep had covered the proto-Earth. The idea that the Earth was once molten goes back to the pioneers of science, including Gottfried Wilhelm Leibniz and the Comte de Buffon, and, of course, was the fundamental assumption of Laplace and his nebular hypothesis. George Darwin, in his fission theory, assumed that the proto-Earth had been molten, allowing a bulge to form and pinch off to become the Moon. In his 1947 article, Daly supposed that at the time of collision, the Earth "was surfaced with a thick layer of liquid." Several authors suggested that the greater early abundance of heat-generating radioactive isotopes, before they had decayed themselves into extinction, could have caused large-scale melting on the proto-Earth, possibly complete melting.[11] The presence of low-density anorthosite in samples from the early Apollo missions suggested that it may have floated to the surface of a lunar magma ocean.[12] Others would extend the idea to the entire solar system, as in a review article from 2011 that concluded: "Theory and observations point to the occurrence of magma ponds or oceans in the early evolution of terrestrial planets and in many early-accreting planetesimals."[13]

As Hartmann and Davis noted in their original article proposing giant impact, "In the case of Earth-sized planets, the models suggest second-largest bodies of 500 to 3000 km radius, and tens of bodies larger than 100 km radius." This means that the giant impact that formed the Moon was only the last of many collisions of the proto-Earth with objects smaller than Theia but still large enough to melt the Earth's outer surface. Indeed, if multiple impacts are likely, then many must have occurred on surfaces that were already molten. An arriving Theia thus likely would have crashed not onto solid rock but onto an ocean of liquid magma. Four scientists explored this possibility in a 2019 article titled "Terrestrial Magma Ocean Origin of the Moon."[14] The lead author is giant-impact specialist Natsuki Hosono, assistant professor in the Graduate School of Science of Kobe University.

One of the deficiencies of SPH is that it does not work well where large differences exist in density across boundaries, as between, say, the core and mantle of a proto-planet. Hosono and colleagues modified their SPH code to better resolve those density changes and found that impact onto a magma ocean would have caused as much as 80 percent of the Moon to come from the proto-Earth, whose isotopic composition would then have dominated, helping to explain the identities. Figure 15.2 is an example of one of Hosono's simulation runs. It illustrates the amazing level of detail from an SPH simulation.

* * *

Given the success of the computer simulations in explaining the origin of the Moon, it was natural for scientists to apply the method to other moons of

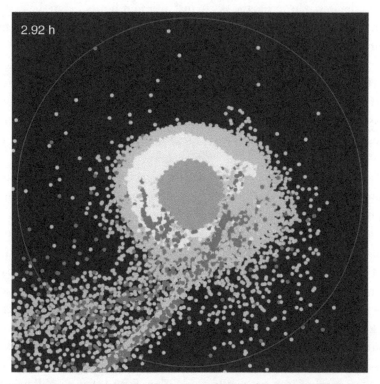

Figure 15.2. Magma ocean simulation, 2.92 hours after impact. Dark gray = target core. Orange = target mantle. Red = target magma ocean. Light gray = impactor core. Blue = impactor mantle. (Hosono et al., "Terrestrial Magma Ocean.")

the solar system, to see whether giant impact could also explain them. One example is the puzzling pair of Martian mini-moons, Phobos and Deimos. It was not clear that the impact model would work for them, since they are so tiny compared with their primary planet; the Moon has about 1 percent of the mass of the Earth, but Phobos and Deimos combined have only about 0.000002 percent of the mass of Mars. Phobos, the son of Ares (Mars), was the Greek god who represented fear, giving us the word "phobia." His brother Deimos personified terror. We met these moons earlier when scientists in the 1870s began to question the nebular hypothesis. The problem was that Phobos, the innermost of the pair, revolves around Mars about three times as fast as Mars rotates, the opposite of what the great hypothesis predicted. Phobos is also closer to Mars than any other known moon to its primary, though so many moons continue to be found that any such statement may soon be superseded. The first astronaut to stand on Mars would see Phobos rise in the west, zip across the sky in just over four hours, set in the east, and then repeat the trip, all in the same Martian day. Deimos takes about thirty hours to go around Mars and is much farther away.

Scientists can get an idea of the composition of bodies like the Martian pair from their calculated density and light spectra and reflectance. On those criteria, both Phobos and Deimos resemble certain classes of asteroids. As shown in Figure 15.3, Phobos is heavily cratered. These are thought to be ancient features, perhaps in the range of 4 billion years old.

The similarity of the two moonlets to objects in the main asteroid belt led to the hypothesis that both represent captured asteroids. On the other hand, both have nearly circular orbits, more consistent with an origin in an impact-generated equatorial disk. The two ideas are equivalent to the capture and co-accretion theories for the origin of the Moon, which giant impact superseded. Could giant impact also explain the origin of Phobos and Deimos? Canup and Julien Salmon took up that question in a 2018 article.[15] The most obvious way to launch a disk into orbit is through impact and we know that impact created the large basins on Mars. But those impacts would have produced disks many times larger than necessary to form the two tiny moons. Canup and Salmon used a modified version of SPH to find out what kind of impacts would work. They found that an oblique strike on Mars by an object about one-thousandth its size could create two small moons, one with the rapid speed of Phobos. Critical data on the pair are scheduled to arrive in 2024 with the Martian Moons Exploration of the Japanese Aerospace Exploration Agency, which will land on Phobos and bring back samples.

Figure 15.3. Phobos with crater Stickney at lower right. (Wikimedia Commons.)

But Phobos and Deimos were not the only objects to be investigated as possible products of giant impact. Far out on the edge of the solar system in the Kuiper belt, a doughnut-shaped ring of icy bodies around the Sun, lies Pluto. For many years, astronomers considered it to be the ninth planet; then, in 2006, the International Astronomical Union downgraded Pluto's status to "dwarf planet." The IAU defines a planet by two properties. First, it must have enough mass to allow its own gravity to compress it into a roughly spherical shape. Second, it must have enough gravitational attraction to sweep up and assimilate nearby objects or else throw them out of the way. Pluto passes the first test but not the second. It has five moons, the largest of which is Charon, discovered only in 1977, which, with a radius of 606 kilometers, is in the size range of known asteroids. Charon has about one-eighth the mass of Pluto, an even greater ratio than our Moon has to the Earth. Scientists consider Charon a "planetoid," not a true moon.

In 2005, Canup used SPH to study the formation of Pluto and Charon, finding giant impact to be a "quite plausible" explanation.[16] Charon might have been born intact, or it could have accreted from a disk thrown off Pluto. Canup thought that giant impact of objects in the range of 1,000 kilometers diameter would have been common in the early Kuiper belt.

In 2015, the NASA New Horizons spacecraft flew near Pluto and came within 27,000 kilometers of Charon. The remote-sensing data showed that the densities of the pair differed by less than 10 percent, suggesting that Pluto and Charon might have been two similar bodies that collided before orbiting each other as the Earth and the Moon do.

Scientists continue their simulations of giant impact, now with a new kind of evidence at hand from the steadily increasing number of exoplanets that astronomers have found around other stars. In February 2019, an article appeared in *Nature* that described the giant impact of two planets ten times the size of the Earth in a faraway solar system named Kepler-107 (see Figure 15.4).[17]

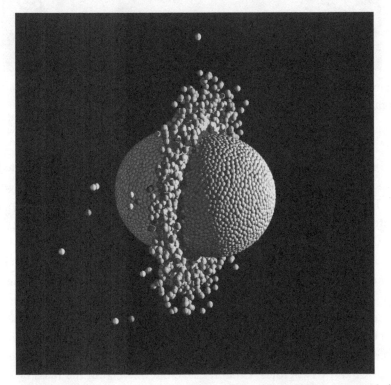

Figure 15.4. SPH simulation of the collision of twin planets. (Bonomo et al., "A Giant Impact.")

16

Summing Up and Looking Ahead

As of the early 2020s, lunar scientists have reached a broad consensus that, as Erik Asphaug of Arizona's Lunar and Planetary Institute summed up, "Earth formed in a series of giant impacts, and the last one made the Moon."[1] To reach that conclusion took nearly half a century after giant impact was introduced. During those many decades, the answer to the great question of the Moon's origin defied the best scientific minds and the latest in scientific instrumentation and computation. The quest may remind us of the search for the Higgs boson, which took forty years and the construction of CERN's Large Hadron Collider, one of the most expensive and complex instruments ever invented. Yet the Moon is by far the most obvious heavenly body, and the Higgs field is present everywhere in the universe. Obvious and omnipresent does not mean easy to understand. Yet in both cases, scientists never gave up, driven by the innate curiosity that brought them to science in the first place.

As for the Moon's craters, there was never any direct evidence that they were volcanic and much evidence against it, as G. K. Gilbert admirably summed up. It simply never made sense. After Gilbert wrote, Alfred Wegener, Robert Dietz, Ralph Baldwin, Reginald Daly, and Eugene Shoemaker, a veritable who's who of earth and planetary science, in turn endorsed impact cratering on the Moon and on the Earth. But minds were made up and they were largely ignored. Then, seventy-one years after Gilbert's paper, Ranger 7 crashed into the northern rim of Mare Nubium, and these scientists were shown to have been right all along. Two other great theories, continental drift and anthropogenic global warming, followed the same path: five decades of rejection despite there being no evidence against them and ample evidence in their favor.[2] Science is not supposed to work this way, but given human nature, it does. Harold Urey and others had argued that going to the Moon would teach us not only about its origin and history but also about those of the Earth and the other planets as well. This has proved to be true in spades; from studying the Moon rocks collected by the Apollo astronauts, we now

Unlocking the Moon's Secrets. James Lawrence Powell, Oxford University Press. © James Lawrence Powell 2023.
DOI: 10.1093/oso/9780197694862.003.0017

know that in the early solar system, many huge bodies like Theia roamed but were swallowed up, leaving us only the infinitesimal isotopes as clues to their one-time existence.

As scientists began from the 1950s onward to use radiometric dating to measure rock ages, it was natural for them to try to find the oldest. Claire Patterson's measurement using lead isotopes in meteorites (including Canyon Diablo) showed that the Earth is 4.5 billion years old, but try as they might, scientists could find no terrestrial rock older than around 3.8 billion years. Geologists assumed that this 700-million-year gap represented the time it took the proto-Earth to cool and solidify from its molten beginning, since only solid rocks would retain the daughter atoms needed for dating.

NASA selected the landing sites for the last three Apollo missions—15, 16, and 17—because of their nearness to the immense Imbrium, Nectaris, and Serenitatis Basins, respectively. When scientists dated the specimens returned by those missions, they found that all had about the same age: around 3.95 billion years. This indicated an intense period of meteorite infall, when an unusually large number of asteroids struck the Moon. In that case, the four inner planets—Mercury, Venus, Earth, and Mars—would have suffered the same barrage as the Moon. This hypothesis became known as the Moon's Late Heavy Bombardment, but it quickly became controversial and the jury is still out. For one thing, high-resolution maps from the Lunar Reconnaissance Orbiter showed rays of debris from Imbrium extending over the other two basins. (In his talk at the 1964 conference on lunar problems, Baldwin had said that the Imbrium impact had deposited "a mantle of debris that buried more than half the visible hemisphere of the Moon.")[3] Thus, it may be that the samples collected in Nectaris and Serenitatis were splashed there from Imbrium, leaving it the only basin dated. The Late Heavy Bombardment remains an area of active research, but certain conclusions seem likely to stand. The Orientale Basin shown earlier in Figure 12.1 is also around 3.8 billion years old, so that together with Imbrium, at least two 1,000-kilometer basins formed on the Moon at around the same time. If we extrapolate from these two to the Earth, considering our planet's greater gravitational attraction for incoming objects, there should have been about forty similar-sized impacts on Earth in the same short time interval. Thus, around 4 billion years ago, dozens of impact basins as large as Orientale and Imbrium must have formed on Earth. As one pair of authors noted, "Now that is a late heavy bombardment!"[4] If the Earth was left with a magma ocean in the wake of the giant impact that created the Moon, this colossal barrage would have ensured that it persisted for hundreds of millions of years.

Apollo 17, the last mission to the Moon with a human crew, took place in December 1972. At the time, most scientists and the public assumed that the program would continue, and indeed, three more missions were in the original plan for Apollo. The proximate reason for the ending of the program in 1972 was that Congress cut NASA's budget. That may have simply reflected the fact that an expensive program created largely to beat the Russians to the Moon had lost political and public support once it succeeded. NASA had other programs to launch, including Skylab, the first space station, and the docking of American and Soviet spacecraft.

Now, fifty years later, the spirit of space exploration and discovery that motivated Apollo has once again arisen. President Barack Obama increased NASA's budget, and in December 2017, President Donald Trump signed Space Policy Directive 1, which called for landing humans on the Moon, followed by crewed missions to Mars. In May 2019, NASA administrator James Bridenstine announced that the new lunar program would be named after Artemis, the Greek god Apollo's twin sister. It would be a collaboration between NASA and private spaceflight companies. NASA announced that the program would be international and would land the first woman and the first person of color on the Moon. Though the Apollo missions focused especially on the second half of Kennedy's charge—"and return him safely to Earth"—Artemis would establish a permanent base on the Moon. Eventually, it would be used to extract resources and to launch spacecraft to Mars.

Eight entities—in chronological order, the Soviet Union, the United States, Japan, the European Space Agency, China, India, Luxembourg, and Israel—have sent spacecraft to the Moon. Others are also planning Moon missions, including South Korea, the United Arab Emirates, and possibly Russia. The most successful to date may have been the Chinese program. In January 2019, Chang'e 4 made the first soft landing on the farside of the Moon. Chang'e 5 landed in Oceanus Procellarum in December 2020 and collected and sent back to Earth samples weighing 1,731 grams. Analysis of these specimens is already bearing fruit, as an article in 2022 reported the presence of water in samples of lunar soil.[5] This was a critical finding, as, among other things, water can be split into hydrogen and vital oxygen. Many other space missions for scientific purposes are planned, heralding a new golden age of space exploration.

* * *

When Shoemaker wrote in 1977 that "impact of solid bodies is the most fundamental process that has taken place on the terrestrial planets"—this at a time before giant impact had caught on—he was more right than he could have known.[6] The word "fundamental" comes from the Latin *fundare*, to found. It is no exaggeration to say that impact founded the solar system, including the planets, moons, comets, and asteroids—and, of course, our Moon in particular. But think how easily it could have turned out differently. Had Theia had a slightly different composition, traveled at a slightly different velocity, or struck the proto-Earth at a slightly different angle, our planet and our Moon would not have resulted. We know that another contingent extraterrestrial event, the asteroid impact that exterminated the dinosaurs and 70 percent of all species, spared a squirrel-sized mammal—our distant ancestor.[7] Now we know that in the random violence of the early solar system, a much larger contingent event took place, one that left a planet on which life could arise, become intelligent, and figure out how lucky it was.

Now, when we look at the lovely, glowing face of our seemingly peaceful heavenly companion, in the back of our minds we hold the knowledge that the Moon was born in violence on a scale that we cannot comprehend. For a host of reasons, we can only be thankful that it survived and remains our one and only Moon.

Acknowledgments

I thank my indefatigable agent, John Thornton, and for expert editorial assistance, Jonathan Cobb.

Notes

Chapter 1

1. J. Needham, *Science and Civilisation in China*, Vol. 3, *Mathematics and the Sciences of the Heavens and the Earth* (Cambridge: Cambridge University Press, 1959), 415–416. https://books.google.com/books?id=jfQ9E0u4pLAC.
2. Karl Popper, *Conjectures and Refutations: The Growth of Scientific Knowledge* (Routledge, 1969), 138.
3. Patricia Curd, "Anaxagoras," in *The Stanford Encyclopedia of Philosophy*, n.d., https://plato.stanford.edu/archives/win2019/entries/anaxagoras/.
4. David Warmflash, "An Ancient Greek Philosopher Was Exiled for Claiming the Moon Was a Rock, Not a God | Science| Smithsonian Magazine," June 20, 1919, https://www.smithsonianmag.com/science-nature/ancient-greek-philosopher-was-exiled-claiming-moon-was-rock-not-god-180972447/#.

Chapter 2

1. Galileo Galilei, *Sidereus Nuncius, or The Sidereal Messenger* (Chicago: University of Chicago Press, 2016).
2. Isaac Disraeli, *Curiosities of Literature, Second Series* (Paris: Baudy's European Library, 1835), https://archive.org/details/disraelicuriosit03disr/mode/2up?q=Galileo.
3. Galileo Galilei, *Dialogue Concerning the Two Chief World Systems, Ptolemaic and Copernican*, trans. Stillman Drake (New York: Modern Library, 2001), 70. https://books.google.com/books?id=c-nIrKjBqOwC.
4. Galileo, *Sidereus Nuncius*, 7.
5. Galileo, *Sidereus Nuncius*, 7.
6. Stephen G. Brush, *A History of Modern Planetary Physics: Nebulous Earth* (Cambridge: Cambridge University Press, 1996), 93.
7. Johannes Kepler, *Kepler's Somnium: The Dream, or Posthumous Work on Lunar Astronomy*, trans. Edward Rosen (Mineola, NY: Dover, 2003), https://books.google.com/books?id=OdCJAS0eQ64C.
8. George Basalla, *Civilized Life in the Universe: Scientists on Intelligent Extraterrestrials* (New York: Oxford University Press, 2006).
9. Robert Hooke, *Micrographia* [1665], Project Gutenberg ebook, http://www.gutenberg.org/files/15491/15491-h/15491-h.htm.

Chapter 3

1. William Sheehan and Richard Baum, "Observations and Inference: Johann Hieronymous Schroeter, 1745–1816," *Journal of the British Astronomical Association* 105 (1995): 171–175.
2. James Sime, *William Herschel and His Work* (Edinburgh: T. & T. Clark, 1900).
3. Sheehan and Baum, "Observations and Inference," 172.
4. Ewen A. Whitaker, *Mapping and Naming the Moon: A History of Lunar Cartography and Nomenclature* (Cambridge: Cambridge University Press, 2003).
5. William Graves Hoyt, *Coon Mountain Controversies: Meteor Crater and the Development of Impact Theory* (Tucson: University of Arizona Press, 1987).
6. "The Great Moon Hoax of 1835," http://hoaxes.org/text/display/the_great_moon_ho ax_of_1835_text/P1.
7. Hoyt, *Coon Mountain Controversies*.
8. "Caroline Herschel," Britannica, https://www.britannica.com/biography/Caroline-Lucretia-Herschel.
9. Grzegorz Racki and Christian Koeberl, "In Search of Historical Roots of the Meteorite Impact Theory: Franz von Paula Gruithuisen as the First Proponent of an Impact Cratering Model for the Moon in the 1820s," *Meteoritics & Planetary Science* 54, no. 10 (2019): 2600–2630, https://doi.org/10.1111/maps.13280.
10. Thomas Hockey et al., eds., *Biographical Encyclopedia of Astronomers* (New York: Springer, 2014), https://doi.org/10.1007/978-1-4419-9917-7; P. M. Ryves, "Mary Adela Blagg," *Obituary Notices, Royal Astronomical Society*, no. 2 (1945): 65–66, https://articles.adsabs.harvard.edu/cgi-bin/nph-iarticle_qu ery?1945MNRAS.105R..65.&defaultprint=YES&filetype=.pdf.
11. Ryves, "Mary Adele Blagg."
12. Hockey et al., *Biographical Encyclopedia*.
13. Agnes M. Clerke, *A Popular History of Astronomy during the Nineteenth Century. History of Astronomy* (London: A. and C. Black, 1902), 265. https://catalog.hathitrust. org/Record/001475525.

Chapter 4

1. James Dwight Dana, "On the Condition of Vesuvius in July, 1834." *American Journal of Science and Arts* 27 (1835): 281–288, https://search.proquest.com/openview/bc335 cd1afd1314d62dfdc0b11e6a0d5/1?pq-origsite=gscholar&cbl=42401.
2. James Dwight Dana, "On the Volcanoes of the Moon," *American Journal of Science and Arts* 2 (second series), no. 6 (1846): 335.
3. James Dwight Dana, *Manual of Geology: Treating of the Principles of the Science with Special Reference to American Geological History* (American book Company, 1895), 11.
4. Dana, "On the Volcanoes of the Moon," 343–345.

5. Aboriginal cultures in Australia knew of meteorites and their possible relation to craters. But there was no way for this knowledge to reach Western scientists. Duane W. Hamacher, "Recorded Accounts of Meteoritic Events in the Oral Traditions of Indigenous Australians," *Archaeoastronomy* 15 (2013): 99–111, https://doi.org/10.48550/arXiv.1408.6368.

6. Ursula B. Marvin, "Ernst Florens Friedrich Chladni (1756–1827) and the Origins of Modern Meteorite Research," *Meteoritics & Planetary Science* 31, no. 5 (1996): 545–588.

7. Marvin, "Ernst Florens Friedrich Chladni."

8. Racki and Koeberl, "In Search of Historical Roots."

9. Richard A. Proctor, *The Moon, Her Motions, Aspect, Scenery, and Physical Condition* (London: Longmans, Green, 1878), https://catalog.hathitrust.org/Record/001991 093; James Nasmyth and James Carpenter, *The Moon: Considered as a Planet, a World, and a Satellite* (London: J. Murray, 1874), http://archive.org/details/moonconsidered a00carpgoog.

10. "Proctor, Richard Anthony," *Complete Dictionary of Scientific Biography*, https://www.encyclopedia.com/science/dictionaries-thesauruses-pictures-and-press-relea ses/proctor-richard-anthony.

11. "Chladni, Ernst Florenz Friedrich." In *Complete Dictionary of Scientific Biography*, accessed April 27, 2023, https://www.encyclopedia.com/science/dictionaries-thes auruses-pictures-and-press-releases/chladni-ernst-florenz-friedrich.

12. Richard Anthony Proctor, *The Moon: Her Motions, Aspect, Scenery, and Physical Condition* (Longmans, Green, 1873), 343.

13. Proctor, *The Moon*, 345.

14. Richard A. Proctor, *The Poetry of Astronomy: A Series of Familiar Essays on the Heavenly Bodies, Regarded Less in Their Strictly Scientific Aspect Than as Suggesting Thoughts Respecting Infinities of Time and Space, of Varitey, of Vitality, and of Development* (Philadelphia: J. B. Lippincott, 1881).

15. Grove Karl Gilbert, "The Moon's Face: A Study of the Origin of Its Features," *Bulletin of the Philosophical Society of Washington* 12 (1893): 241–292.

Chapter 5

1. S. J. Pyne, *Grove Karl Gilbert: A Great Engine of Research* (Iowa City: University of Iowa Press, 2007), https://books.google.com/books?id=BjYObIHgITgC.

2. James Lawrence Powell, *Grand Canyon: Solving Earth's Grandest Puzzle* (New York: Penguin, 2006), https://books.google.com/books?id=bVqdezfrqdUC.

3. William Morris Davis, "Grove Karl Gilbert," *Biographical Memoirs* (Washington, DC: National Academy of Sciences, 1927).

4. William Morris Davis, "Biographical Memoir of Grove Karl Gilbert," *Biographical Memoirs of Members of the National Academy of Sciences* 21, no. 5 (1927): 176.

5. G. K. Gilbert, *The Moon's Face: A Study of the Origin of Its Features* (Washington, DC: Philosophical Society of Washington, 1893).

6. William Morris Davis, "Lunar Craters," *The Nation* 56, no. 1454 (1893): 342–343.
7. Davis, "Grove Karl Gilbert."
8. A. E. Foote, "Geological Features of the Meteoric Iron Locality in Arizona," *Proceedings of the Academy of Natural Sciences of Philadelphia* 43 (1891): 407.
9. Davis, "Grove Karl Gilbert."
10. G. K. Gilbert, "The Inculcation of Scientific Method by Example: With an Illustration Drawn from the Quaternary Geology of Utah," *American Journal of Science* 31 (1886): 288
11. G. K. Gilbert, "The Origin of Hypotheses, Illustrated by the Discussion of a Topographic Problem," *Science* 3, no. 53 (1896): 1–13.
12. Davis, "Grove Karl Gilbert."
13. Grove Karl Gilbert, " The Inculcation of Scientific Method by Example: With an Illustration Drawn from the Quaternary Geology of Utah," *American Journal of Science* 31, no. 184 (1886): 12.
14. James Lawrence Powell, *Four Revolutions in the Earth Sciences; from Heresy to Truth* (New York: Columbia University Press, 2015).
15. Hoyt, *Coon Mountain Controversies*.
16. Davis, "Grove Karl Gilbert."
17. Grzegorz Racki et al., "Ernst Julius Öpik's (1916) Note on the Theory of Explosion Cratering on the Moon's Surface: The Complex Case of a Long-Overlooked Benchmark Paper," *Meteoritics & Planetary Science* 49, no. 10 (2014): 1851–1874.
18. Ralph B. Baldwin, *The Measure of the Moon* (Chicago: University of Chicago Press, 1963).
19. Racki, Grzegorz, Christian Koeberl, Tõnu Viik, Elena A. Jagt-Yazykova, and John W. M. Jagt, "Ernst Julius Öpik's (1916) Note on the Theory of Explosion Cratering on the Moon's Surface—The Complex Case of a Long-Overlooked Benchmark Paper," *Meteoritics & Planetary Science* 49, no. 10 (2014): 1851–1874.
20. Herbert E. Ives, "Some Large-Scale Experiments Imitating the Craters of the Moon," *Astrophysical Journal* 50 (November 1919): 245, https://doi.org/10.1086/142503.
21. Oliver E. Buckley and Karl K. Darrow, "Herbert Eugene Ives (1882–1953)," *Biographical Memoirs* (Washington, DC: National Academy of Sciences,1956), 145–189, http://www.nasonline.org/publications/biographical-memoirs/memoir-pdfs/ives-herbert.pdf.
22. Alfred Wegener, "Die Entstehung der Mondkrater," *Science of Nature* 9, no. 30 (1921): 592–594.
23. B. Barringer, "Historical Notes on the Odessa Meteorite Crater," *Meteoritics* 3, no. 4 (December 1967): 161, http://adsabs.harvard.edu/abs/1967Metic...3..161B.

Chapter 6

1. L. J. Spencer, "Meteorite Craters as Topographical Features on the Earth's Surface," *Geographical Journal* 81, no. 3 (1933): 227–243, https://doi.org/10.2307/1784038.

2. Walter Bucher, "Geology of Jeptha Knob," Kentucky Geological Survey, 1920, https://catalog.hathitrust.org/Record/000531821.

3. W. H. Bradley, "Walter Herman Bucher," *Biographical Memoirs* (Washington, DC: National Academy of Sciences, 1969), 19–34, http://www.nasonline.org/publications/biographical-memoirs/memoir-pdfs/bucher-walter-h.pdf.

4. Bucher, "Geology of Jeptha Knob."

5. Walter H. Bucher, "Cryptovolcanic Structures in the United States," *Proceedings of the 16th International Geological Congress* 2 (1933): 1055–1084.

6. K. Mark, *Meteorite Craters* (Tucson: University of Arizona Press, 1995), 153, https://books.google.com/books?id=nRex5I86vH4C.

7. John D. Boon and Claude C. Albritton Jr., "Meteorite Craters and Their Possible Relationship to 'Cryptovolcanic Structures,'" *Field and Laboratory* 5, no. 1 (1936): 1–9.

8. John D. Boon and Claude C. Albritton Jr, "Meteorite Scars in Ancient Rocks," *Field and Laboratory* 5, no. 2 (1937): 5.

9. Walter H. Bucher, "Cryptoexplosion Structures Caused from without or from within the Earth? ('Astroblemes' or 'Geoblemes'?)," *American Journal of Science* 261, no. 7 (1963): 597–649.

10. J. Bourgeois and S. Koppes, "Robert S. Dietz and the Recognition of Impact Structures on Earth," *Earth Sciences History* 17, no. 2 (1998): 139–156.

11. Powell, *Four Revolutions.*

12. Robert S. Dietz, "Earth, Sea, and Sky: Life and Times of a Journeyman Geologist," *Annual Review of Earth and Planetary Sciences* 22 (1994): 1–32.

13. Dietz, "Earth, Sea, and Sky." 21.

14. Dietz, "Earth, Sea, and Sky," 21.

15. Ralph B. Baldwin, "The Meteoritic Origin of Lunar Craters, with Plate III," *Popular Astronomy* 50 (August 1942): 356.

16. Robert S. Dietz, "The Meteoritic Impact Origin of the Moon's Surface Features," *The Journal of Geology* 54, no. 6 (1946): 361.

17. Ralph B. Baldwin, *The Face of the Moon* (Chicago: University of Chicago Press, 1949).

18. Carlyle Smith Beals, M. J. S. Innes, and J. A. Rottenberg, "The Search for Fossil Meteorite Craters—I," *Current Science* 29, no. 6 (1960): 205–218.

19. Ursula B. Marvin, "Oral Histories in Meteoritics and Planetary Science: X. Ralph B. Baldwin," *Meteoritics & Planetary Science* 38, no. S7 (July 2003): A163–A175, https://doi.org/10.1111/j.1945-5100.2003.tb00326.x.

Chapter 7

1. D. H. Levy, *Shoemaker by Levy: The Man Who Made an Impact.* Princeton: Princeton University Press, 2002, https://books.google.com/books?id=NQBPj1s0G1wC; Susan W. Kieffer, "Eugene M. Shoemaker, 1928–1997," *Biographical Memoirs* (Washington, DC: National Academy of Sciences, 2015).

2. Eugene M. Shoemaker, "Impact Mechanics at Meteor Crater, Arizona," US Geological Survey, 1959, http://pubs.er.usgs.gov/publication/ofr59108.

3. J. Green, "Hookes and Spurrs in Selenology," *Annals of the New York Academy of Sciences* 123, no. 2 (1965): 373–402, https://doi.org/10.1111/j.1749-6632.1965.tb20 376.x.

4. Kurd von Bülow, "Proof of the Volcanic Origin of Most Lunar Craters and of Tectonic Maria," *Annals of the New York Academy of Sciences* 123, no. 2 (1965): 528–531, https://doi.org/10.1111/j.1749-6632.1965.tb20384.x.

5. Walter H. Bucher, "The Largest So-Called Meteorite Scars in Three Continents as Demonstrably Tied to Major Terrestrial Structures," *Annals of the New York Academy of Sciences* 123, no. 2 (1965): 897–903, https://doi.org/10.1111/j.1749-6632.1965.tb20 408.x.

6. John F. Kennedy, "Excerpt from the 'Special Message to the Congress on Urgent National Needs," NASA, http://www.nasa.gov/vision/space/features/jfk_speech_t ext.html.

7. Levy, *Shoemaker by Levy*.

8. Wegener, "Die Entstehung Der Mondkrater."

9. NASA, "NASA FACTS Volume 2 Number 6 PROJECT RANGER," n.d., https://en.wik isource.org/wiki/NASA_FACTS_Volume_2_number_6_PROJECT_RANGER.

10. "Says Two Crops a Day Grow on the Moon," *New York Times*, October 9, 1921, https:// timesmachine.nytimes.com/timesmachine/1921/10/09/107027707.pdf.

11. Richard Feynman, *"Surely You're Joking, Mr. Feynman!": Adventures of a Curious Character* (New York: Norton, 1985), 343.

Chapter 8

1. "Laplace, Marquis Pierre Simon de," *A Dictionary of Scientists* (New York: Oxford University Press, 1999), https://www.oxfordreference.com/view/10.1093/acref/ 9780192800862.001.0001/acref-9780192800862-e-849.

2. Stephen G. Brush, *A History of Modern Planetary Physics: Nebulous Earth* (Cambridge University Press, 1996), 22.

3. Brush, *A History of Modern Planetary Physics*, 15.

4. "Edouard Albert Roche," Oxford Reference," https://www.oxfordreference.com/ view/10.1093/oi/authority.20110803100425308.

5. Daniel Kirkwood, "The Satellites of Mars and the Nebular Hypothesis," *The Observatory* 1 (1878): 280–282.

6. Daniel Kirkwood, "The Cosmogony of Laplace," *Proceedings of the American Philosophical Society* 18, no. 104 (1879): 324–326.

7. Brush, *A History of Modern Planetary Physics*.

8. G. Brent Dalrymple, *The Age of the Earth* (Stanford: Stanford University Press, 1991). Dalrymple has an excellent treatment of George Darwin's work.

9. Dalrymple, *The Age of the Earth*.

10. O. Fisher, "On the Physical Cause of the Ocean Basins," *Nature* 25 (1882): 243–244, https://doi.org/10.1038/025243a0.

11. Thomas Jefferson Jackson See, *Researches on the Evolution of the Stellar Systems: The Capture Theory of Cosmical Evolution* (Lynn, MA: T. P. Nichols, 1896), http://hdl.handle.net/2027/mdp.39015023544219.

12. Thomas Jefferson Jackson See, "On the Cause of the Remarkable Circularity of the Orbits of the Planets and Satellites and on the Origin of the Planetary System," *Popular Astronomy* 17 (1909): 263-272.

13. F. R. Moulton, "An Attempt to Test the Nebular Hypothesis by an Appeal to the Laws of Dynamics," *The Astrophysical Journal* 11 (1900): 103.

Chapter 9

1. Kennedy, "Excerpt."

2. J. R. Arnold, Jacob Bigeleisen, and Clyde Hutchison, "Harold Clayton Urey, 1893–1981," *Biographical Memoirs* (Washington, DC: National Academy of Sciences, 1995), 363–411, http://www.nasonline.org/publications/biographical-memoirs/memoir-pdfs/urey-harold.pdf.

3. Quoted in Matthew Shindell, "Harold C. Urey: Science, Religion, and Cold War Chemistry," Science History Institute, January 2014, https://www.sciencehistory.org/distillations/harold-c-urey-science-religion-and-cold-war-chemistry.

4. Marvin, "Oral Histories."

5. Arnold, Bigeleisen, and Hutchison, "Harold Clayton Urey."

6. Stephen G. Brush, "Nickel for Your Thoughts: Urey and the Origin of the Moon," *Science* 217, no. 4563 (1982): 897.

7. Powell, *Four Revolutions*; James Lawrence Powell, "Premature Rejection in Science: The Case of the Younger Dryas Impact Hypothesis," *Science Progress* 105, no. 1 (January 2022): 00368504211064272, https://doi.org/10.1177/00368504211064272.

8. Harold C. Urey, "The Origin and Development of the Earth and Other Terrestrial Planets," *Geochimica et Cosmochimica Acta* 1, no. 4 (January 1951): 209–277, https://doi.org/10.1016/0016-7037(51)90001-4.

9. Loyd S. Swenson Jr., James M. Grimwood, and Charles C. Alexander, *This New Ocean* (Washington, DC: NASA, 1989), 294–297, https://history.nasa.gov/SP-4201/cover.htm.

10. "A New Look at Copernicus," *Time*, December 9, 1966, Science, https://content.time.com/time/subscriber/article/0,33009,898477,00.html.

11. William K. Hartmann, "Paleocratering of the Moon: Review of Post-Apollo Data," *Astrophysics and Space Science* 17, no. 1 (1972): 48–64.

12. D. E. Wilhelms, *To a Rocky Moon: A Geologist's History of Lunar Exploration* (Tucson: University of Arizona Press, 1993), https://books.google.com/books?id=Cn_vAAAAMAAJ.

13. Kenneth Chang, "Neil Armstrong: First Man on the Moon, and Its First Great Geologist," *New York Times*, July 6, 2019, Science, https://www.nytimes.com/2019/07/06/science/neil-armstrong-moon-rocks.html.

Chapter 10

1. Lunar Sample Preliminary Examination Team, "Preliminary Examination of Lunar Samples from Apollo 11," *Science* 165, no. 3899 (1969): 1211–1227.
2. New discoveries have revealed that the Moon does have detectable amounts of water. It is present near the poles in areas that are in permanent shadow and also in minute amounts in the Moon's infinitesimal atmosphere. It may have been delivered by impacting comets and asteroids or produced by protons from the Sun interacting with oxygen-bearing minerals. See "Is There an Atmosphere on the Moon?," NASA, April 12, 2013, https://www.nasa.gov/mission_pages/LADEE/news/lunar-atmosphere. html#.VCzkF2eSxQA.
3. "Orientale" means "eastern." Mare Orientale is on the eastern side of the Moon as seen from Earth but on the western side to an astronaut on the Moon. In 1961, recognizing what was to come, the International Astronomical Union adopted the astronaut's perspective.
4. Ian I. Mitroff, *The Subjective Side of Science: A Philosophical Inquiry Into the Psychology of the Apollo Moon Scientists* (Netherlands: Elsevier Scientific Publishing Company, 1974).
5. Robert Clayton and Toshiko Mayeda, "Genetic Relations between the Moon and Meteorites," *Proceedings of the 6th Lunar Science Conference* (1975): 1761–1769, http://articles.adsabs.harvard.edu/pdf/1975LPSC....6.1761C.
6. An inscribed monument or "stele" discovered in Egypt in 1799. It contained the same text in hieroglyphics, Demotic, and Greek and was key to the deciphering of Egyptian writing.

Chapter 11

1. Eric Hand, "Recent Volcanic Eruptions on the Moon," *Science*, October 2014, https://www.science.org/content/article/recent-volcanic-eruptions-moon.

Chapter 12

1. Joseph A. Burns, "Guest Editorial," *Icarus* 24, no. 4 (April 1975): 3, https://doi.org/10.1016/0019-1035(75)90054-8.
2. William K. Hartmann, "The Giant Impact Hypothesis: Past, Present (and Future?)," *Philosophical Transactions of the Royal Society A: Mathematical, Physical and Engineering Sciences* 372, no. 2024 (August 2014): 20130249–20130249, https://doi.org/10.1098/rsta.2013.0249.
3. William K. Hartmann and Donald R. Davis, "Satellite-Sized Planetesimals and Lunar Origin," *Icarus* 24, no. 4 (April 1975): 504–515, https://doi.org/10.1016/0019-1035(75)90070-6.

4. Hartmann, "Paleocratering."
5. "Help Name Five Newly Discovered Moons of Jupiter!," Carnegie Institution for Science, February 2019, https://carnegiescience.edu/NameJupitersMoons.
6. Powell, *Four Revolutions*.
7. Henry Cooper, "Letter from the Space Station," *The New Yorker*, June 8, 1987
8. Another great scientist, Gerry Wasserburg, told an amusing story about the imposing Cameron, who in 1975 gave a talk on "The Origin of the Solar System" at Caltech. He began with the Sun and proto-planetary disk and wound up with the gas giants, the rocky terrestrial planets, and the Moon. "The audience sat in awed silence until someone asked, 'What did you do on the seventh day? [Cameron] responded, 'I rested.'" Gerald J. Wasserburg, "Alistair Graham Walter Cameron," *Physics Today* 59, no. 1 (January 2006): 68, https://physicstoday.scitation.org/doi/10.1063/1.2180186.

Chapter 13

1. Reginald A Daly, "Origin of the Moon and Its Topography," *Proceedings of the American Philosophical Society* 90, no. 2 (May 1946): 104–119.
2. Powell, *Four Revolutions*.
3. Quoted in Cooper, "Letter from the Space Station."
4. Wilhelms, *To a Rocky Moon*.
5. Hartmann, "The Giant Impact Hypothesis."
6. William K. Hartmann, "Moon Origin: The Impact-Trigger Hypothesis," in *Origin of the Moon*, ed. William K. Hartmann, Roger Phillips, and G. Jeffrey Taylor (Houston: Lunar & Planetary Institute, 1986), 579–608.
7. Richard A. Kerr, "Making the Moon from a Big Splash," *Science* 226 (1984): 1060–1062.
8. Powell, "Premature Rejection."
9. David Walker and James F. Hays, "Plagioclase Flotation and Lunar Crust Formation," *Geology* 5 (1977): 425–428.
10. John A. Wood, "Moon over Mauna Loa: A Review of Hypotheses of Formation of Earth's Moon," in *Origin of the Moon*, ed. William K. Hartmann, Roger Phillips, and G. Jeffrey Taylor (Houston: Lunar & Planetary Institute, 1986), 17–55.
11. Kerr, "Making the Moon."

Chapter 14

1. A. N. Halliday, "The Origin of the Moon," *Science* 338, no. 6110 (2012): 1040–1041.
2. U. Wiechert et al., "Oxygen Isotopes and the Moon-Forming Giant Impact," *Science* 294, no. 5541 (October 2001): 345–348, https://doi.org/10.1126/science.1063037.
3. Edward D. Young et al., "Oxygen Isotopic Evidence for Vigorous Mixing during the Moon-Forming Giant Impact," *Science* 351, no. 6272 (January 2016): 493–496, https://doi.org/10.1126/science.aad0525.

4. Robert N. Clayton and Toshiko K. Mayeda, "Oxygen Isotope Studies of Achondrites," *Geochimica et Cosmochimica Acta* 60, no. 11 (June 1996): 1999–2017, https://doi.org/10.1016/0016-7037(96)00074-9.

5. Jérôme Gattacceca et al., "The Meteoritical Bulletin, No. 108," *Meteoritics & Planetary Science* 55, no. 5 (May 2020): 1146–1150, https://doi.org/10.1111/maps.13493.

6. Tim Elliott and Sarah T. Stewart, "Shadows Cast on Moon's Origin," *Nature* 504, no. 7478 (December 2013): 90–91, https://doi.org/10.1038/504090a.

7. Erick J. Cano, Zachary D. Sharp, and Charles K. Shearer, "Distinct Oxygen Isotope Compositions of the Earth and Moon," *Nature Geoscience* 13 (March 2020): 270–274, https://doi.org/10.1038/s41561-020-0550-0.

8. Rebecca A. Fischer, Nicholas G. Zube, and Francis Nimmo, "The Origin of the Moon's Earth-like Tungsten Isotopic Composition from Dynamical and Geochemical Modeling," *Nature Communications* 12, no. 1 (2021): 1–6.

Chapter 15

1. Daniel J. Price, "Smoothed Particle Hydrodynamics and Magnetohydrodynamics," *Journal of Computational Physics* 231, no. 3 (February 2012): 759–794, https://doi.org/10.1016/j.jcp.2010.12.011.

2. W. Benz, W. L. Slattery, and A. G. W. Cameron, "The Origin of the Moon and the Single-Impact Hypothesis I," *Icarus* 66, no. 3 (June 1986): 515–535, https://doi.org/10.1016/0019-1035(86)90088-6.

3. The figure is from Horton E. Newsom and Stuart Ross Taylor, "Geochemical Implications of the Formation of the Moon by a Single Giant Impact," *Nature* 338, no. 6210 (March 1989): 29–34, https://doi.org/10.1038/338029a0.

4. Robin M. Canup and Erik Asphaug, "Origin of the Moon in a Giant Impact Near the End of the Earth's Formation," *Nature* 412, no. 6848 (2001): 708.

5. Ed Finn, "Robin Canup," *Popular Science*, March 2019, https://www.popsci.com/scitech/article/2004-10/robin-canup/.

6. Robin M. Canup, "Dynamics of Lunar Formation," *Annual Review of Astronomy and Astrophysics* 42 (2004): 441–475.

7. M. Ćuk and S. T. Stewart, "Making the Moon from a Fast-Spinning Earth: A Giant Impact Followed by Resonant Despinning," *Science* 338, no. 6110 (November 2012): 1047–1052, https://doi.org/10.1126/science.1225542.

8. Stewart has a YouTube video illustrating evection resonance at https://youtu.be/35tO6K04HDo.

9. R. M. Canup, "Forming a Moon with an Earth-like Composition via a Giant Impact," *Science* 338, no. 6110 (November 2012): 1052–1055, https://doi.org/10.1126/science.1226073.

10. H. J. Melosh, "An Isotopic Crisis for the Giant Impact Origin of the Moon?," 72nd Annual Meeting of the Meteoritical Society, Nancy, France, 2009, https://www.lpi.usra.edu/meetings/metsoc2009/pdf/5104.pdf.

11. Harold C. Urey, "The Cosmic Abundances of Potassium, Uranium, and Thorium and the Heat Balances of the Earth, the Moon, and Mars," *Proceedings of the National Academy of Sciences* 41, no. 3 (March 1955): 127–144, https://doi.org/10.1073/pnas.41.3.127.

12. David Walker, "Lunar and Terrestrial Crust Formation," *Journal of Geophysical Research* 88, no. S01 (1983): B17, https://doi.org/10.1029/JB088iS01p00B17.

13. Linda T. Elkins-Tanton, "Magma Oceans in the Inner Solar System," *Annual Review of Earth and Planetary Sciences* 40, no. 1 (May 2012): 113–139, https://doi.org/10.1146/annurev-earth-042711-105503.

14. Natsuki Hosono et al., "Terrestrial Magma Ocean Origin of the Moon," *Nature Geoscience* 12, no. 6 (June 2019): 418–423, https://doi.org/10.1038/s41561-019-0354-2.

15. Robin Canup and Julien Salmon, "Origin of Phobos and Deimos by the Impact of a Vesta-to-Ceres Sized Body with Mars," *Science Advances* 4, no. 4 (April 2018): eaar6887, https://doi.org/10.1126/sciadv.aar6887.

16. Robin M. Canup, "A Giant Impact Origin of Pluto-Charon," *Science* 307, no. 5709 (January 2005): 546–550, https://doi.org/10.1126/science.1106818.

17. Aldo S. Bonomo et al., "A Giant Impact as the Likely Origin of Different Twins in the Kepler-107 Exoplanet System," *Nature Astronomy* 3, no. 5 (May 2019): 416, https://doi.org/10.1038/s41550-018-0684-9.

Chapter 16

1. Erik Asphaug, "Impact Origin of the Moon?," *Annual Review of Earth and Planetary Sciences* 42, no. 1 (May 2014): 551–578, https://doi.org/10.1146/annurev-earth-050212-124057.

2. Powell, *Four Revolutions*.

3. Ralph B. Baldwin, "The Origin of Lunar Features," *Annals of the New York Academy of Sciences* 123, no. 2 (July 1965): 543–554, https://doi.org/10.1111/j.1749-6632.1965.tb20386.x.

4. William F. Bottke and Marc D. Norman, "The Late Heavy Bombardment," *Annual Review of Earth and Planetary Sciences* 45 (2017): 619–647.

5. Jianjun Liu et al., "Evidence of Water on the Lunar Surface from Chang'e-5 in-Situ Spectra and Returned Samples," *Nature Communications* 13, no. 1 (June 2022): 3119, https://doi.org/10.1038/s41467-022-30807-5.

6. Eugene M. Shoemaker, "Why Study Impact Craters," in *Impact and Explosion Cratering: Planetary and Terrestrial Implications*, ed. D. J. Roddy, R. O. Pepin, and R. B. Merrill (New York: Pergamon, 1977), 1–10.

7. James Lawrence Powell, *Night Comes to the Cretaceous: Dinosaur Extinction and the Transformation of Modern Geology* (New York: W. H. Freeman, 1998).

Bibliography

Arnold, J. R., Jacob Bigeleisen, and Clyde Hutchison. "Harold Clayton Urey, 1893–1981." *Biographical Memoirs*, 363–411. Washington, DC: National Academy of Sciences, 1995. http://www.nasonline.org/publications/biographical-memoirs/memoir-pdfs/urey-harold.pdf.

Asphaug, Erik. "Impact Origin of the Moon?" *Annual Review of Earth and Planetary Sciences* 42, no. 1 (May 2014): 551–578. https://doi.org/10.1146/annurev-earth-050212-124057.

Baldwin, Ralph. B. *The Face of the Moon.* Chicago: University of Chicago Press, 1949.

Baldwin, Ralph B. *The Measure of the Moon.* Chicago: University of Chicago Press, 1963.

Baldwin, Ralph B. "The Meteoritic Origin of Lunar Craters, with Plate III." *Popular Astronomy* 50 (August 1942): 356.

Baldwin, Ralph B. "The Origin of Lunar Features." *Annals of the New York Academy of Sciences* 123, no. 2 (July 1965): 543–554. https://doi.org/10.1111/j.1749-6632.1965.tb20386.x.

Barringer, B. "Historical Notes on the Odessa Meteorite Crater." *Meteoritics* 3, no. 4 (December 1967): 161. http://adsabs.harvard.edu/abs/1967Metic...3..161B.

Basalla, George. *Civilized Life in the Universe: Scientists on Intelligent Extraterrestrials.* New York: Oxford University Press, 2006.

Beals, Carlyle Smith, M. J. S. Innes, and J. A. Rottenberg. "The Search for Fossil Meteorite Craters—I." *Current Science* 29, no. 6 (1960): 205–218.

Beer, W., and J. H. Maedler. *Der Mond Nach Seinen Kosmischen Und Individuellen Verhältnissen Oder Allgemeine Vergleichende Selenographie.* Simon Schropp, 1837. https://books.google.com/books?id=FcFLAAAAcAAJ.

Benz, W., W. L. Slattery, and A. G. W. Cameron. "The Origin of the Moon and the Single-Impact Hypothesis I." *Icarus* 66, no. 3 (June 1986): 515–535. https://doi.org/10.1016/0019-1035(86)90088-6.

Bonomo, Aldo S., Li Zeng, Mario Damasso, Zoë M. Leinhardt, Anders B. Justesen, Eric Lopez, Mikkel N. Lund, et al. "A Giant Impact as the Likely Origin of Different Twins in the Kepler-107 Exoplanet System." *Nature Astronomy* 3, no. 5 (May 2019): 416. https://doi.org/10.1038/s41550-018-0684-9.

Boon, John D., and Claude C. Albritton Jr. "Meteorite Craters and Their Possible Relationship to 'Cryptovolcanic Structures.'" *Field and Laboratory* 5, no. 1 (1936): 1–9.

Boon, John D., and Claude C. Albritton Jr. "Meteorite Scars in Ancient Rocks." *Field and Laboratory* 5, no. 2 (1937): 5.

Bottke, W. "Interpreting the Elliptical Crater Populations on Mars, Venus, and the Moon." *Icarus* 145, no. 1 (May 2000): 108–121. https://doi.org/10.1006/icar.1999.6323.

Bourgeois, J., and S. Koppes. "Robert S. Dietz and the Recognition of Impact Structures on Earth." *Earth Sciences History* 17, no. 2 (1998): 139–156.

Bradley, W. H. "Walter Hermann Bucher." *Biographical Memoirs*, 19–34. Washington, DC: National Academy of Sciences, 1969. http://www.nasonline.org/publications/biographical-memoirs/memoir-pdfs/bucher-walter-h.pdf.

Brush, Stephen G. "Nickel for Your Thoughts: Urey and the Origin of the Moon." *Science* 217, no. 4563 (1982): 897.

Brush, Stephen G. *A History of Modern Planetary Physics: Nebulous Earth*. Cambridge: Cambridge University Press, 1996.

Bucher, Walter H. "Cryptoexplosion Structures Caused from without or from within the Earth? ('Astroblemes' or 'Geoblemes'?)." *American Journal of Science* 261, no. 7 (1963): 597–649.

Bucher, Walter H. "Cryptovolcanic Structures in the United States." *Proceedings of the 16th International Geological Congress,* 2 (1933): 1055–1084.

Bucher, Walter H. "Geology of Jeptha Knob." Kentucky Geological Survey, 1920. https://catalog.hathitrust.org/Record/000531821.

Bucher, Walter H. "The Largest So-Called Meteorite Scars in Three Continents as Demonstrably Tied to Major Terrestrial Structures." *Annals of the New York Academy of Sciences* 123, no. 2 (1965): 897–903. https://doi.org/10.1111/j.1749-6632.1965.tb20408.x.

Buckley, Oliver E., and Karl K. Darrow. "Herbert Eugene Ives (1882–1953)." *Biographical Memoirs*, 145–189. Washington, DC: National Academy of Sciences, 1956. http://www.nasonline.org/publications/biographical-memoirs/memoir-pdfs/ives-herbert.pdf.

Bülow, Kurd von. "Proof of the Volcanic Origin of Most Lunar Craters and of Tectonic Maria." *Annals of the New York Academy of Sciences* 123, no. 2 (1965): 528–531. https://doi.org/10.1111/j.1749-6632.1965.tb20384.x.

Burns, Joseph A. "Guest Editorial." *Icarus* 24, no. 4 (April 1975): 3. https://doi.org/10.1016/0019-1035(75)90054-8.

Cano, Erick J., Zachary D. Sharp, and Charles K. Shearer. "Distinct Oxygen Isotope Compositions of the Earth and Moon." *Nature Geoscience* 13 (March 2020): 270–274. https://doi.org/10.1038/s41561-020-0550-0.

Canup, Robin M. "Dynamics of Lunar Formation." *Annual Review of Astronomy and Astrophysics* 42 (2004): 441–475.

Canup, Robin M. "Forming a Moon with an Earth-like Composition via a Giant Impact." *Science* 338, no. 6110 (November 2012): 1052–1055. https://doi.org/10.1126/science.1226073.

Canup, Robin M. "A Giant Impact Origin of Pluto-Charon." *Science* 307, no. 5709 (January 2005): 546–550. https://doi.org/10.1126/science.1106818.

Canup, Robin M., and Erik Asphaug. "Origin of the Moon in a Giant Impact Near the End of the Earth's Formation." *Nature* 412, no. 6848 (2001): 708.

Canup, Robin, and Julien Salmon. "Origin of Phobos and Deimos by the Impact of a Vesta-to-Ceres Sized Body with Mars." *Science Advances* 4, no. 4 (April 2018): eaar6887. https://doi.org/10.1126/sciadv.aar6887.

"Caroline Herschel." Britannica. https://www.britannica.com/biography/Caroline-Lucretia-Herschel.

Chang, Kenneth. "Neil Armstrong: First Man on the Moon, and Its First Great Geologist." *New York Times*, July 6, 2019, Science. https://www.nytimes.com/2019/07/06/science/neil-armstrong-moon-rocks.html.

Chladni, E. F. F. *Uber Den Ursprung Der von Pallas Gefindenen Und Anderer Ihr Ahnlicher Eisenmassen, Und Uber Einige Darnit in Verbindung Stehende Naturerscheinungen.* Riga: Johann Friedrich Hartknoch, 1794.

"Chladni, Ernst Florenz Friedrich." In *Complete Dictionary of Scientific Biography.* Accessed April 27, 2023. https://www.encyclopedia.com/science/dictionaries-thes auruses-pictures-and-press-releases/chladni-ernst florenz-friedrich.

Clayton, Robert N., and Toshiko K. Mayeda. "Genetic Relations between the Moon and Meteorites." *Proceedings of the 6th Lunar Science Conference* (1975): 1761–1769. http://articles.adsabs.harvard.edu/pdf/1975LPSC....6.1761C.

Clayton, Robert N., and Toshiko K. Mayeda. "Oxygen Isotope Studies of Achondrites." *Geochimica et Cosmochimica Acta* 60, no. 11 (June 1996): 1999–2017. https://doi.org/10.1016/0016-7037(96)00074-9.

Cooper, Henry. "Letter from the Space Station." *The New Yorker*, June 8, 1987.

Ćuk, M., and S. T. Stewart. "Making the Moon from a Fast-Spinning Earth: A Giant Impact Followed by Resonant Despinning." *Science* 338, no. 6110 (November 2012): 1047–1052. https://doi.org/10.1126/science.1225542.

Curd, Patricia. "Anaxagoras." In *The Stanford Encyclopedia of Philosophy*, n.d. https://plato.stanford.edu/archives/win2019/entries/anaxagoras/.

Dalrymple, G. Brent. *The Age of the Earth.* Stanford: Stanford University Press, 1991.

Daly, Reginald A. "Origin of the Moon and Its Topography." *Proceedings of the American Philosophical Society* 90, no. 2 (May 1946): 104–119.

Dana, James Dwight. "On the Condition of Vesuvius in July, 1834." *American Journal of Science and Arts* 27 (1835): 281–288. https://search.proquest.com/openview/bc335 cd1afd1314d62dfdc0b11e6a0d5/1?pq-origsite=gscholar&cbl=42401.

Dana, James Dwight. "On the Volcanoes of the Moon." *American Journal of Science and Arts* 2 (second series), no. 6 (1846): 352–353.

Dana, James Dwight. *Manual of Geology: Treating of the Principles of the Science with Special Reference to American Geological History.* American Book Company, 1895.

Davis, William Morris. "Grove Karl Gilbert." *Biographical Memoirs.* Washington, DC: National Academy of Sciences, 1927.

Davis, William Morris. "Lunar Craters." *The Nation* 56, no. 1454 (1893): 342–343.

Dietz, Robert S. "Earth, Sea, and Sky: Life and Times of a Journeyman Geologist." *Annual Review of Earth and Planetary Sciences* 22 (1994): 1–32.

Disraeli, Isaac. *Curiosities of Literature, Second Series.* Paris: Baudy's European Library, 1835. https://archive.org/details/disraelicuriosit03disr/mode/2up?q=Galileo.

"Edouard Albert Roche." Oxford Reference.https://www.oxfordreference.com/view/10.1093/oi/authority.20110803100425308.

Elkins-Tanton, Linda T. "Magma Oceans in the Inner Solar System." *Annual Review of Earth and Planetary Sciences* 40, no. 1 (May 2012): 113–139. https://doi.org/10.1146/annurev-earth-042711-105503.

Elliott, Tim, and Sarah T. Stewart. "Shadows Cast on Moon's Origin." *Nature* 504, no. 7478 (December 2013): 90–91. https://doi.org/10.1038/504090a.

Feynman, Richard. *"Surely You're Joking, Mr. Feynman!": Adventures of a Curious Character.* New York: Norton, 1985.

Finn, Ed. "Robin Canup." *Popular Science*, March 2019. https://www.popsci.com/scitech/article/2004-10/robin-canup/.

Fischer, Rebecca A., Nicholas G. Zube, and Francis Nimmo. "The Origin of the Moon's Earth-like Tungsten Isotopic Composition from Dynamical and Geochemical Modeling." *Nature Communications* 12, no. 1 (2021): 1–6.

Fisher, O. "On the Physical Cause of the Ocean Basins." *Nature* 25 (1882): 243–244. https://doi.org/10.1038/025243a0.

Foote, A. E. "Geological Features of the Meteoric Iron Locality in Arizona." *Proceedings of the Academy of Natural Sciences of Philadelphia* 43 (1891): 407.

Galileo Galilei. *Dialogue Concerning the Two Chief World Systems, Ptolemaic and Copernican.* Translated by Stillman Drake. New York: Modern Library, 2001. https://books.google.com/books?id=c-nIrKjBqOwC.

Galileo Galilei. *Sidereus Nuncius, or The Sidereal Messenger.* Chicago: University of Chicago Press, 2016.

Gattacceca, Jérôme, Francis M. McCubbin, Audrey Bouvier, and Jeffrey N. Grossman. "The Meteoritical Bulletin, No. 108." *Meteoritics & Planetary Science* 55, no. 5 (May 2020): 1146–1150. https://doi.org/10.1111/maps.13493.

Gilbert, G. K. "The Inculcation of Scientific Method by Example: With an Illustration Drawn from the Quaternary Geology of Utah." *American Journal of Science* 31 (1886): 288.

Gilbert, G. K. *The Moon's Face: A Study of the Origin of Its Features.* Washington, DC: Philosophical Society of Washington, 1893.

Gilbert, G. K. "The Origin of Hypotheses, Illustrated by the Discussion of a Topographic Problem." *Science* 3, no. 53 (1896): 1–13.

"The Great Moon Hoax of 1835." http://hoaxes.org/text/display/the_great_moon_hoax_of_1835_text/P1.

Green, J. "Hookes and Spurrs in Selenology." *Annals of the New York Academy of Sciences* 123, no. 2 (1965): 373–402. https://doi.org/10.1111/j.1749-6632.1965.tb20376.x.

Halliday, A. N. "The Origin of the Moon." *Science* 338, no. 6110 (2012): 1040–1041.

Hamacher, Duane W. "Recorded Accounts of Meteoritic Events in the Oral Traditions of Indigenous Australians," *Archaeoastronomy* 15 (2013): 99–111. https://doi.org/10.48550/arXiv.1408.6368.

Hand, Eric. "Recent Volcanic Eruptions on the Moon." *Science*, October 2014. https://www.science.org/content/article/recent-volcanic-eruptions-moon.

Hartmann, William K. "The Giant Impact Hypothesis: Past, Present (and Future?)." *Philosophical Transactions of the Royal Society A: Mathematical, Physical and Engineering Sciences* 372, no. 2024 (August 2014): 20130249–20130249. https://doi.org/10.1098/rsta.2013.0249.

Hartmann, William K. "Moon Origin: The Impact-Trigger Hypothesis." In *Origin of the Moon,* edited by William K. Hartmann, Roger Phillips, and G. Jeffrey Taylor, 579–608. Houston: Lunar & Planetary Institute, 1986.

Hartmann, William K. "Paleocratering of the Moon: Review of Post-Apollo Data." *Astrophysics and Space Science* 17, no. 1 (1972): 48–64.

Hartmann, William K., and Donald R. Davis. "Satellite-Sized Planetesimals and Lunar Origin." *Icarus* 24, no. 4 (April 1975): 504–515. https://doi.org/10.1016/0019-1035(75)90070-6.

"Help Name Five Newly Discovered Moons of Jupiter!" Carnegie Institution for Science, February 2019. https://carnegiescience.edu/NameJupitersMoons.

Hockey, Thomas, Virginia Trimble, Thomas R. Williams, Katherine Bracher, Richard A. Jarrell, Jordan D. Marché, JoAnn Palmeri, and Daniel W. E. Green, eds. *Biographical*

Encyclopedia of Astronomers. New York: Springer, 2014. https://doi.org/10.1007/978-1-4419-9917-7.

Hooke, Robert. *Micrographia* [1665]. Project Gutenberg ebook. http://www.gutenberg.org/files/15491/15491-h/15491-h.htm.

Hosono, Natsuki, Shun-ichiro Karato, Junichiro Makino, and Takayuki R. Saitoh. "Terrestrial Magma Ocean Origin of the Moon." *Nature Geoscience* 12, no. 6 (June 2019): 418–423. https://doi.org/10.1038/s41561-019-0354-2.

Hoyt, William Graves. *Coon Mountain Controversies: Meteor Crater and the Development of Impact Theory*. Tucson: University of Arizona Press, 1987.

"Is There an Atmosphere on the Moon?" NASA, April 12, 2013. https://www.nasa.gov/mission_pages/LADEE/news/lunar-atmosphere.html#.VCzkF2eSxQA.

Ives, Herbert E. "Some Large-Scale Experiments Imitating the Craters of the Moon." *Astrophysical Journal* 50 (November 1919): 245. https://doi.org/10.1086/142503.

Kennedy, John F. "Excerpt from the 'Special Message to the Congress on Urgent National Needs.'" NASA. http://www.nasa.gov/vision/space/features/jfk_speech_text.html.

Kepler, Johannes. *Kepler's Somnium: The Dream, Or Posthumous Work on Lunar Astronomy*. Translated by Edward Rosen. Mineola, NY: Dover, 2003. https://books.google.com/books?id=OdCJAS0eQ64C.

Kerr, Richard A. "Making the Moon from a Big Splash." *Science* 226 (1984): 1060–1062.

Kieffer, Susan W. "Eugene M. Shoemaker, 1928–1997." *Biographical Memoirs*. Washington, DC: National Academy of Sciences, 2015.

Kirkwood, Daniel. "The Cosmogony of Laplace." *Proceedings of the American Philosophical Society* 18, no. 104 (1879): 324–326.

Kirkwood, Daniel. "The Satellites of Mars and the Nebular Hypothesis." *The Observatory* 1 (1878): 280–282.

"Laplace, Marquis Pierre Simon de." *A Dictionary of Scientists*. New York: Oxford University Press, 1999. https://www.oxfordreference.com/view/10.1093/acref/9780192800862.001.0001/acref-9780192800862-e-849.

Levy, D. H. *Shoemaker by Levy: The Man Who Made an Impact*. Princeton: Princeton University Press, 2002. https://books.google.com/books?id=NQBPj1s0G1wC.

Liu, Jianjun, Bin Liu, Xin Ren, Chunlai Li, Rong Shu, Lin Guo, Songzheng Yu, et al. "Evidence of Water on the Lunar Surface from Chang'e-5 in-Situ Spectra and Returned Samples." *Nature Communications* 13, no. 1 (June 2022): 3119. https://doi.org/10.1038/s41467-022-30807-5.

Lunar Sample Preliminary Examination Team. "Preliminary Examination of Lunar Samples from Apollo 11." *Science* 165, no. 3899 (1969): 1211–1227.

Mark, K. *Meteorite Craters*. Tucson: University of Arizona Press, 1995. https://books.google.com/books?id=nRex5I86vH4C.

Marvin, Ursula B. "Ernst Florens Friedrich Chladni (1756–1827) and the Origins of Modern Meteorite Research." *Meteoritics & Planetary Science* 31, no. 5 (1996): 545–588.

Marvin, Ursula B. "Oral Histories in Meteoritics and Planetary Science: X. Ralph B. Baldwin." *Meteoritics & Planetary Science* 38, no. S7 (July 2003): A163–A175. https://doi.org/10.1111/j.1945-5100.2003.tb00326.x.

Melosh, H. J. "An Isotopic Crisis for the Giant Impact Origin of the Moon?" 72nd Annual Meeting of the Meteoritical Society, Nancy, France, 2009. https://www.lpi.usra.edu/meetings/metsoc2009/pdf/5104.pdf.

Mitroff, I. I. *The Subjective Side of Science: A Philosophical Inquiry into the Psychology of the Apollo Moon Scientists*. Amsterdam: Elsevier, 1974. https://books.google.com/books?id=ysraAAAAMAAJ.

Moulton, F. R. "An Attempt to Test the Nebular Hypothesis by an Appeal to the Laws of Dynamics." *The Astrophysical Journal* 11 (1900): 103.

NASA. "NASA FACTS Volume 2 Number 6 PROJECT RANGER." n.d. https://en.wikisou rce.org/wiki/NASA_FACTS_Volume_2_number_6_PROJECT_RANGER.

Nasmyth, James, and James Carpenter. *The Moon: Considered as a Planet, a World, and a Satellite*. London: J. Murray, 1874. http://archive.org/details/moonconsidereda00c arpgoog.

Needham, J. *Science and Civilisation in China*, Vol. 3, *Mathematics and the Sciences of the Heavens and the Earth*. Cambridge: Cambridge University Press, 1959. https://books. google.com/books?id=jfQ9E0u4pLAC.

"A New Look at Copernicus." *Time*, December 9, 1966, Science. https://content.time.com/time/subscriber/article/0,33009,898477,00.html.

Newsom, Horton E., and Stuart Ross Taylor. "Geochemical Implications of the Formation of the Moon by a Single Giant Impact." *Nature* 338, no. 6210 (March 1989): 29–34. https://doi.org/10.1038/338029a0.

Pickering, W. H. "The Origin of the Lunar Formations." *Publications of the Astronomical Society of the Pacific* 32 (April 1, 1920): 116. https://doi.org/10.1086/122944.

Powell, James Lawrence. *Four Revolutions in the Earth Sciences; from Heresy to Truth*. New York: Columbia University Press, 2015.

Powell, James Lawrence. *Grand Canyon: Solving Earth's Grandest Puzzle*. New York: Penguin, 2006. https://books.google.com/books?id=bVqdezfrqdUC.

Powell, James Lawrence. *Night Comes to the Cretaceous: Dinosaur Extinction and the Transformation of Modern Geology*. New York: W. H. Freeman, 1998.

Powell, James Lawrence. "Premature Rejection in Science: The Case of the Younger Dryas Impact Hypothesis." *Science Progress* 105, no. 1 (January 2022): 00368504211064272. https://doi.org/10.1177/00368504211064272.

Price, Daniel J. "Smoothed Particle Hydrodynamics and Magnetohydrodynamics." *Journal of Computational Physics* 231, no. 3 (February 2012): 759–794. https://doi.org/10.1016/j.jcp.2010.12.011.

Proctor, Richard Anthony. *The Moon: Her Motions, Aspect, Scenery, and Physical Condition*. Longmans, Green, 1873.

Proctor, Richard A. *The Moon, Her Motions, Aspect, Scenery, and Physical Condition*. London: Longmans, Green, 1878. https://catalog.hathitrust.org/Record/001991093.

Proctor, Richard A. *The Poetry of Astronomy: A Series of Familiar Essays on the Heavenly Bodies, Regarded Less in Their Strictly Scientific Aspect Than as Suggesting Thoughts Respecting Infinities of Time and Space, of Varitey, of Vitality, and of Development*. Philadelphia: J. B. Lippincott, 1881.

"Proctor, Richard Anthony." *Complete Dictionary of Scientific Biography*. https://www.encyclopedia.com/science/dictionaries-thesauruses-pictures-and-press-releases/proc tor-richard-anthony.

Pyne, S. J. *Grove Karl Gilbert: A Great Engine of Research*. Iowa City: University of Iowa Press, 2007. https://books.google.com/books?id=BjYObIHgITgC.

Racki, Grzegorz, and Christian Koeberl. "In Search of Historical Roots of the Meteorite Impact Theory: Franz von Paula Gruithuisen as the First Proponent of an Impact

Cratering Model for the Moon in the 1820s." *Meteoritics & Planetary Science* 54, no. 10 (2019): 2600–2630. https://doi.org/10.1111/maps.13280.

Racki, Grzegorz, Christian Koeberl, Tõnu Viik, Elena A. Jagt-Yazykova, and John W. M. Jagt. "Ernst Julius Öpik's (1916) Note on the Theory of Explosion Cratering on the Moon's Surface: The Complex Case of a Long-Overlooked Benchmark Paper." *Meteoritics & Planetary Science* 49, no. 10 (2014): 1851–1874.

Ryves, P.M. "Mary Adele Blagg." *Obituary Notices, Royal Astronomical Society*, no. 2 (1945): 65–66. https://articles.adsabs.harvard.edu/cgi-bin/nph-iarticle_query?1945MNRAS.105R..65.&defaultprint=YES&filetype=.pdf.

"Says Two Crops a Day Grow on the Moon." *New York Times*, October 9, 1921. https://timesmachine.nytimes.com/timesmachine/1921/10/09/107027707.pdf.

See, Thomas Jefferson Jackson. "On the Cause of the Remarkable Circularity of the Orbits of the Planets and Satellites and on the Origin of the Planetary System." *Popular Astronomy* 17 (1909): 263–272.

See, Thomas Jefferson Jackson. *Researches on the Evolution of the Stellar Systems: The Capture Theory of Cosmical Evolution.* Lynn, MA: T. P. Nichols, 1896. http://hdl.handle.net/2027/mdp.39015023544219.

Sheehan, William, and Richard Baum. "Observations and Inference: Johann Hieronymous Schroeter, 1745–1816." *Journal of the British Astronomical Association* 105 (1995): 171–175.

Shindell, Matthew. "Harold C. Urey: Science, Religion, and Cold War Chemistry." Science History Institute, January 2014. https://www.sciencehistory.org/distillations/harold-c-urey-science-religion-and-cold-war-chemistry.

Shoemaker, Eugene M. "Impact Mechanics at Meteor Crater, Arizona." US Geological Survey, 1959. http://pubs.er.usgs.gov/publication/ofr59108.

Shoemaker, Eugene M. "Why Study Impact Craters." In *Impact and Explosion Cratering: Planetary and Terrestrial Implications*, edited by D. J. Roddy, R. O. Pepin, and R. B. Merrill, 1–10. New York: Pergamon, 1977.

Sime, James. *William Herschel and His Work.* Edinburgh: T. & T. Clark, 1900.

Spencer, L. J. "Meteorite Craters as Topographical Features on the Earth's Surface." *Geographical Journal* 81, no. 3 (1933): 227–243. https://doi.org/10.2307/1784038.

Swenson, Loyd S., James M. Grimwood, and Charles C. Alexander. *This New Ocean.* Washington, DC: NASA, 1989. https://history.nasa.gov/SP-4201/cover.htm.

Urey, H. C. *The Planets, Their Origin and Development. Mrs. Hepsa Ely Silliman Memorial Lectures.* Yale University Press, 1952. https://books.google.com/books?id=KMG7AAAAIAAJ.

Urey, Harold C. "The Cosmic Abundances of Potassium, Uranium, and Thorium and the Heat Balances of the Earth, the Moon, and Mars." *Proceedings of the National Academy of Sciences* 41, no. 3 (March 1955): 127–144. https://doi.org/10.1073/pnas.41.3.127.

Urey, Harold C. "The Origin and Development of the Earth and Other Terrestrial Planets." *Geochimica et Cosmochimica Acta* 1, no. 4 (January 1951): 209–277. https://doi.org/10.1016/0016-7037(51)90001-4.

Walker, David. "Lunar and Terrestrial Crust Formation." *Journal of Geophysical Research* 88, no. S01 (1983): B17. https://doi.org/10.1029/JB088iS01p00B17.

Walker, David, and James F. Hays. "Plagioclase Flotation and Lunar Crust Formation." *Geology* 5 (1977): 425–428.

Warmflash, David. "An Ancient Greek Philosopher Was Exiled for Claiming the Moon Was a Rock, Not a God | Science| Smithsonian Magazine," June 20, 1919. https://www.

smithsonianmag.com/science-nature/ancient-greek-philosopher-was-exiled-claim
ing-moon-was-rock-not-god-180972447/#.

Wasserburg, Gerald J. "Alistair Graham Walter Cameron." *Physics Today* 59, no. 1 (January
2006): 68. https://physicstoday.scitation.org/doi/10.1063/1.2180186.

Wegener, Alfred. "Die Entstehung der Mondkrater." *Science of Nature* 9, no. 30
(1921): 592–594.

Whitaker, Ewen A. *Mapping and Naming the Moon: A History of Lunar Cartography and
Nomenclature.* Cambridge: Cambridge University Press, 2003.

Wiechert, U., A. N. Halliday, D.-C. Lee, G. A. Snyder, L. A. Taylor, and D. Rumble.
"Oxygen Isotopes and the Moon-Forming Giant Impact." *Science* 294, no. 5541
(October 2001): 345–348. https://doi.org/10.1126/science.1063037.

Wilhelms, D. E. *To a Rocky Moon: A Geologist's History of Lunar Exploration.*
Tucson: University of Arizona Press, 1993. https://books.google.com/books?id=Cn_
vAAAAMAAJ.

Wood, John A. "Moon over Mauna Loa: A Review of Hypotheses of Formation of Earth's
Moon." In *Origin of the Moon*, edited by William K. Hartmann, Roger Phillips, and
G. Jeffrey Taylor, 17–55. Houston: Lunar & Planetary Institute, 1986.

Young, Edward D., Issaku E. Kohl, Paul H. Warren, David C. Rubie, Seth A. Jacobson, and
Alessandro Morbidelli. "Oxygen Isotopic Evidence for Vigorous Mixing during the
Moon-Forming Giant Impact." *Science* 351, no. 6272 (January 2016): 493–496. https://
doi.org/10.1126/science.aad0525.

Index